NEW UNIVERSE

The Revolution in Our Understanding of the Cosmos

STEPHEN P. MARAN
AND LAURENCE A. MARSCHALL

BENBELLA BOOKS, INC.
Dallas, TX

BENBELLA

BenBella Books, Inc.
6440 N. Central Expressway, Suite 503
Dallas, TX 75206
Send feedback to feedback@benbellabooks.com
www.benbellabooks.com

Printed in the United States of America
10 9 8 7 6 5 4 3 2 1

Library of Congress Cataloging-in-Publication Data

Maran, Stephen P.
  Galileo's new universe : the revolution in our understanding of the cosmos / Stephen P. Maran and Laurence A. Marschall.
      p. cm.
  ISBN 978-1-933771-59-5
  1. Astronomy—History. 2. Astronomy—History--17th century. 3. Galilei, Galileo, 1564-1642. 4. Discoveries in science. I. Marschall, Laurence A. II. Title.

  QB28.M37 2008
  520.9'032—dc22
                                2008045907

Proofreading by Stacia Seaman
Cover design by Laura Watkins
Text design and composition by PerfecType, Nashville, TN
Printed by Bang Printing

Distributed by Perseus Distribution
perseusdistribution.com

To place orders through Perseus Distribution:
Tel: (800) 343-4499
Fax: (800) 351-5073
E-mail: orderentry@perseusbooks.com

Significant discounts for bulk sales are available. Please contact Robyn White at robyn@benbellabooks.com or (214) 750-3600.

TABLE OF CONTENTS

ACKNOWLEDGMENTS

Stephen P. Maran: I am indebted to Prof. John C. Brandt and Dr. Michael J. Mumma for helpful advice and suggestions on observations and physics of comets. I thank my wife, Sally Scott Maran, and our children Michael, Enid, and Elissa for their love and support.

Laurence A. Marschall: Thanks to Owen Gingerich and Donald Goldsmith for much good advice and many good conversations. To Albert Van Helden for sending a copy of his seminal 1977 paper on the invention of the telescope. To all the faculty and staff in the Gettysburg College Physics Department who have supported me over the years. And most of all to my Ellen, Emma, and Geoff, who have been encouraging me to keep on writing.

The authors are especially grateful to their agent, Skip Barker, and to BenBella's Publisher, Glenn Yeffeth, for their encouragement and enabling roles in bringing this project to fruition. We also thank the dedicated staff at BenBella Books for their work in editing, design, and production.

INTRODUCTION

No one is certain who built the first telescope, but we do know who first used it to systematically study the heavens. Sometime around the beginning of the 1600s, reports of a magical tube that had the power to bring distant objects closer began to circulate throughout Europe. Galileo Galilei, a professor of mathematics at the University of Padua, caught word of this curious device during the summer of 1609 and immediately knew he had to have one. Details of its construction were not clear, but Galileo was able to figure out a way to make an instrument that did just what the reports claimed. "First I prepared a tube of lead, at the ends of which I fitted two glass lenses, both plane on one side while on the other side one was spherically convex and the other concave." Looking through the tube, Galileo satisfied himself that distant objects really did seem larger and more distinct. "Finally, sparing neither labor nor expense, I succeeded in constructing for myself so excellent an instrument that objects seen by means of it appeared nearly one thousand times larger and over thirty times closer than when regarded with our natural vision."[1]

---

[1] Galilei, Galileo. *Discoveries and Opinions of Galileo*. Stillman Drake, trans, New York: Anchor, 1957. 29.

In the fall of 1609, Galileo turned his new *perspicillum* skyward—the word telescope was not yet in common use, and Galileo's name for it simply denoted a spyglass or seeing device—and what we knew about the Universe changed overnight. For instance, the Moon, which scholars at the time believed was smooth, uniform, and perfectly spherical, surprisingly revealed a surface that was "uneven, rough, and full of cavities and prominences, being not unlike the face of the earth, relieved by chains of mountains and deep valleys."[2] The Milky Way, thought to be an unbroken cloud of light, resolved itself into a myriad of stars in the far-seeing glass. Galileo's consciousness was jolted by the realization that the Moon and the Earth were two of many worlds in a vast cosmos.

Four hundred years after Galileo, telescopes hundreds of times the diameter of Galileo's crude perspicillum probe the depths of an expanding Universe each night, while spacecraft send back digital images of the heavens seen in infrared and X-rays, types of radiation Galileo never dreamed of. The surface of the Moon now bears the footprints of American astronauts, and lunar geology is a well-developed science. The revolution in understanding the cosmos that started with Galileo has resulted in an ever-accelerating series of discoveries.

To commemorate the 400th anniversary of Galileo's momentous observations, the International Astronomical Union, a global association of research astronomers, declared 2009 the International Year of Astronomy, to

> mark the monumental leap forward that followed Galileo's first use of the telescope for astronomical observations, and portray astronomy as a peaceful global scientific endeavour that unites astronomers in an international, multicultural family of scientists working together to find answers to some of the

---

[2] Ibid., 31.

most fundamental questions that humankind has ever asked.[3]

As members of this community of astronomers, we welcome this opportunity to open our "family" album and examine how Galileo's first indistinct glimpses into space compare with the unbounded cosmos that we study today. Rather than compiling an exhaustive history of the past four centuries—which could easily run into multiple volumes—we have chosen to jump back and forth between the perspective of Galileo's seventeenth-century natural philosophy and the perspective of our twenty-first-century astrophysics so that readers can appreciate better both the profound novelty of the first steps Galileo took and the great distance we have traveled since then. We want to demonstrate the genius of Galileo and, at the same time, highlight the dazzling changes that have resulted from 400 years of progress in astronomy.

Accordingly, each chapter in this book has two parts: first a description of what Galileo and his contemporaries knew (or thought they knew) about a particular celestial object—like Jupiter or the Moon—followed by a review of current astronomical knowledge on the same subject.

It has been said that the more things change, the more they stay the same. That may be true in some realms of human endeavor, but it is seldom true for science. Healthy science is always finding answers to old problems while continuously moving forward along new lines of inquiry. Though some of the questions Galileo faced still puzzle us today, the overall path of scientific development in the past four centuries has been to fill in gaps in knowledge and push our vision outward to the edges of the Universe.

Admittedly, if he were around today, Galileo might find that many human foibles have not altered much in 400 years. Public acceptance for scientific discoveries, in some quarters at least, is still granted only grudgingly, especially when the implications of those

---

[3] "The IAU Announces the IYA." International Astronomical Union. 27 October 2006. <http://www.astronomy2009.org/index.php/?option=com_content&view=article&id=106>

discoveries run counter to some political or religious orthodoxy. In Galileo's time, such troublesome findings were suppressed by bans on speech and writing, and sometimes through corporal or capital punishment. Though we have no Inquisition in the twenty-first century, political appointees and agency "public affairs officers" still threaten government-funded scientists, discouraging speech, testimony, and publication about perceived "sensitive" subjects like global warming. Interest groups sometimes interfere with effective communication of discoveries on topics ranging from stem cell research to the Big Bang theory. Conservative schoolboards require that disclaimers of evolution be placed in biology texts, echoing the Catholic Church's mandate of 1616 that Galileo refer to the Copernican system only as "mere theory." Just as the Inquisition relied on obsolete writings of Greek philosophers to criticize the unconventional findings of Renaissance scientists, contemporary antiscience polemicists cite outdated research and misapplied statistics to oppose scientific positions that they find displeasing.

Nonetheless, we are sure that Galileo would applaud the high degree of public and private support for scientific research in the modern world, just as he would delight at the expanding vistas that we of the twenty-first century enjoy. We share this delight, but even though we know much more than Galileo could have imagined about the Universe, we also marvel at the pioneering work he did four centuries ago. To celebrate the findings of the giant telescopes that scan the heavens today is to honor the seventeenth-century genius who pointed the way.

CHAPTER 1

The Long View: Galileo's Life and Legacy

A BRIEF BIOGRAPHY OF GALILEO

Galileo Galilei, the eldest son of Vincenzo Galilei, was born in Pisa in 1564. He might have been a musician had he followed in the footsteps of his father, an accomplished lute player and composer whose songs and instrumental works are still performed today. Or, had he the inclination, he might have been a doctor, the course of study he (or more likely his father) chose when he enrolled at the University of Pisa at the age of 17. In the end, however, it was the beauty of numbers and geometry that caught his fancy, especially insofar as they applied to the real world. While watching a chandelier in the Cathedral of Pisa, the story goes, young Galileo realized that the time it takes a pendulum of a certain length to swing back and forth is unaffected by the amplitude or extent of the swing. It is said that he used his pulse to time the oscillations. Many years later, near the end of his life, Galileo was to design a clock based on the regularity of a swinging pendulum. (He was, in effect, the father of the grand-father clock.)

Galileo found his vocation at the University of Pisa, but he never managed to complete his degree. It was not because of lack of motivation. The high cost of higher education is one of those things that hasn't changed over the centuries, and since his father earned little money as a musician and had other children to support, the Galileis were unable to come up with the tuition. Failing to win a scholarship that might have kept him in school, Galileo moved back home, but he continued to observe the workings of nature, to tinker, and to write. He displayed such a sharp mind and such skill in devising and building mechanical instruments (his design for a precision balance was particularly elegant) that his work began to attract local attention. In 1589, Ferdinand de' Medici, the Duke of Tuscany, was informed by an advisor that one of his subjects was not sufficiently rewarded for his talents. Ferdinand, who understood that inventive genius was as valuable as gold to a mercantile state, promptly arranged for Galileo's appointment to the faculty of the University of Pisa as a lecturer in mathematics.

Recognition of his abilities was all Galileo needed to make a mark in the world. Quick-witted and ambitious, he stayed in Pisa for only three years before he was offered the chair of mathematics at the University of Padua. It was a prestigious position, and Galileo, over the better part of the next two decades, did some of his most important scientific work there.

The boundaries between professional disciplines in the late 1500s were not as finely drawn as they are today. Though nominally professor of mathematics, Galileo's research and teaching in Padua involved much of what we would today call engineering and physics. He invented a device for measuring temperature, the forerunner of the thermometer, and made extensive observations of the mathematical behavior of moving bodies, a field we now call *kinematics*. He also gave lectures in military engineering and, as an offshoot of these classes, devised a tool called the "geometric and military compass," which was a sort of multipurpose slide rule and measuring device that saved time and effort in operating and targeting artillery.

In the late 1590s, he wrote a manual describing the usefulness of his invention—his first published book—and established a profitable side-business manufacturing and selling the instruments.

We regard Galileo as one of the founding fathers of modern astronomy, but he had done nothing but terrestrial physics and practical engineering up to this time. Still, it is clear that his interests were wide-ranging. He was aware of the controversial Sun-centered system of Copernicus, and he carried on correspondence with many other active scientists of his time, including Johannes Kepler. But it was not until 1609, when he constructed his first telescope, that he made any systematic observations of the heavens.

Galileo's first thoughts about the telescope, even then, were primarily of a practical nature. It would be a very useful tool for military men, he knew, since it would allow them to assess the activities of approaching ships and armies from a distance. As a skilled craftsman, he even entertained the thought of manufacturing and selling his telescopes, which were the best in Europe. After 18 years at Padua, he was tiring of teaching and frustrated at the limited remuneration it provided. On the other hand, he realized that the life of an instrument-maker would tie him down more than the life of a professor.

So Galileo, always creative, eagerly seized on another way to profit from the telescope.He presented one of his first instruments to the Venetian Senate, and in return was granted a request for an increased salary and what amounted to tenure—an assurance of a lifetime professorship at Padua. Had Galileo been satisfied with only financial security at this point, and had he not been curious about what his new instrument might reveal about the world around him, astronomers might not be celebrating his genius today.

Late in 1609, however, Galileo began to look skyward with his telescope and to make discoveries that were to change both his life and the course of science. He saw mountains on the Moon, satellites of Jupiter, phases of Venus, and countless faint stars in the Milky Way—discoveries that form the organizing themes for the later chapters of this book. By early 1610, Galileo was ready to announce his discoveries to the world, in the form of a short book—little more than a pamphlet—titled *Sidereus Nuncius*, or the "The Starry Messenger."

Galileo's book was slim, but its effect was profound. It quickly passed from reader to reader—the seventeenth-century version of mass-media coverage—and within a short time, the obscure professor of mathematics was a celebrity, not just in Tuscany, but throughout all of Europe. Less than five years after its publication, scholars in Peking were discussing its findings with interest: Jesuit missionaries had translated it into Chinese.

More important to Galileo, the little book soon elicited an offer of patronage from a wealthy admirer. Galileo had dedicated the book to Cosimo de' Medici, recent successor to Galileo's old patron, Ferdinand, as ruler of Tuscany. With a nod to continued patronage, Galileo proposed to name the new satellites of Jupiter that he had discovered after the Medici family. Duke Cosimo, flattered and impressed, made Galileo the chief mathematician and philosopher to the Medicis in Florence, relieving Galileo of the teaching and administrative duties of the university and freeing him to write, correspond, and carry out observations and experiments.

While Galileo's fame was virtually immediate, acceptance of his discoveries was not. The telescope was the first optical instrument to reveal sights invisible to the naked eye, and its principles were only beginning to be understood in the early 1600s. (Galileo actually devised an early form of the microscope, too, but microscopes did not come into wide use for another 50 years.) Scientists are by nature skeptical of new claims, and some astronomers suggested that the things Galileo saw were illusions produced in the tube itself—a sort of magician's trick. Others, using inferior instruments, were unable to see for themselves what Galileo had seen.

Those who hesitated to accept Galileo, however, were not just exercising healthy skepticism. The implications of the new discoveries, as we describe in the chapters of this book, contrasted so sharply with accepted principles of physics and cherished theological beliefs, that it seemed almost heresy to embrace them without further investigation.

Immediately following the publication of *Sidereus Nuncius*, Galileo sent telescopes around Europe so that others, using instruments of suitable quality, could view the sights he described with their own eyes. In 1611, he went to Rome and demonstrated the instrument to Jesuit mathematicians there. The visit was a great success, and Galileo was feted with feasts and honorific dinner speeches. He was made member of one of the first scientific organizations, the Accademia-dei Lincei ("Academy of the Lynx-eyed"), a society of learned men who corresponded regularly about new discoveries and new ideas of the time. (Galileo was the sixth member of that select group.) It was during Galileo's initiation into the Lincei that Greek poet and theologian John Demisiani lyrically referred to the new invention as a "telescope"—a far-seeing device—the name it has borne ever since.

Galileo's fame was assured, but ironically, his period of astronomical discovery was relatively short. Between 1610 and 1613, he carried on systematic observations of the heavens, making important discoveries about Venus, the Sun, and Saturn. But after that, he used the telescope only sporadically. In part, this was because he ran out of things to study. There were myriads of lights in the sky, but the ones he wrote about first were the easiest targets. The other planets—Mercury and Mars—and all of the stars, were featureless blobs of light, not much different through the telescope than to the unaided eye.

By 1615, Galileo was past 50 years old, and he suffered increasingly from physical ailments. Yet his mind was as sharp as ever, and until his death in 1642, he was an active correspondent, polemicist, and writer. His observations with the telescope had convinced him of the absolute truth of Copernicus's system, a Universe in which all the planets orbited the Sun. He presented the evidence for a Sun-centered Universe with such clarity and forcefulness that he soon raised the hackles of more conservative clergy and academics, who held to the ancient belief that the Earth was the center of all creation. Galileo was beginning to be regarded as a threat to the established order, a dangerous state of affairs in the days of the Roman Inquisition.

Aware of growing controversy, Galileo went to Rome in 1616 to try to rally support for his point of view. Though there was considerable respect for his intelligence and integrity among the Church fathers, Galileo was unable to prevail. Naturally argumentative, he did not suffer fools gladly, a trait he shared with his father, who had written (in a book on music theory): "It appears to me that those who try to prove an assertion by relying simply on the weight of authority act very absurdly." Cardinal Robert Bellarmine, acting for the Inquisition, admonished Galileo that henceforth he was not to teach or to advocate the doctrine of a moving Earth or a Sun-centered Universe. Galileo left Rome overtly chastened and obedient, though in his heart he knew he was right.

The injunction not to teach Copernicanism ultimately led Galileo into great difficulty with the Church, but despite his precarious position, Galileo managed to write three of the most important scientific works of the seventeenth century during the last 25 years of his life. Spirited, lucid, and entertaining, these books are important milestones on the road to modern science.

The first, *The Assayer*, published in 1623, was a response to criticism he faced, not on Copernicanism, but on comets. The book did not advance our knowledge of the subject—Galileo was dead wrong in maintaining that comets are meteorological rather than astronomical phenomena. But it is a powerful statement of the role of observation, experiment, and mathematics in the understanding of the physical world. Science, wrote Galileo, "is written in this grand book, the Universe. . . . It is written in the language of mathematics, and its characters are triangles, circles, and other geometric figures. . . ." This view, which we take for granted today, was revolutionary at the time. It set the agenda for the next four centuries of science and inspired thinkers like Newton, Einstein, and Heisenberg to look for the mathematical laws which govern the cosmos. When scientists today speak of discovering a Theory of Everything, they are following in the footsteps of Galileo.

The second great book of Galileo, which brought him into direct conflict with the Church, was the *Dialogue on the Great World Systems*, written in the decade following *The Assayer* and published in 1632. It is Galileo's ultimate defense of Copernicanism, thinly disguised as a conversation on the pros and cons of both the Sun-centered and the Earth-centered Universe. Though Galileo believed he had the support of the Pope to publish it, his obvious bias immediately brought matters to a head, and Galileo was summoned before the Inquisition. Tried for heresy in 1633, he was forced to renounce all claims that the Earth moved, and his book was withdrawn from circulation. For the remainder of his life, Galileo remained under house arrest at his home near Florence.

Unable to teach Copernicanism, Galileo spent his latter years consolidating his work on the motion of bodies and the nature of matter. He published a final book on this subject, *Two New Sciences*, in 1638, four years before his death. The laws of motion he set down there pointed the way to an explanation of the most vexing question his telescopic observations had raised: what was it that made the planets go around the Sun? Isaac Newton, directly built on Galileo's work in his 1687 *Mathematical Principles of Natural Philosophy*, in which he set down the mathematical laws that governed the motion of everything in the Universe, from apples to planets. Newton once wrote that if he saw further than others, it was because he was standing on the shoulders of giants. Foremost among them, without a doubt, was Galileo.

Anyone can go out today and buy a telescope that will show the satellites of Jupiter, the craters on the Moon, and the phases of Venus at least as clearly as Galileo's first telescope, so it is tempting to think that anyone with a well-designed instrument in the seventeenth century could have discovered what Galileo did. Perhaps that is true, but Galileo did much more than build good telescopes; he used them to their best advantage. His systematic observations, which we describe in the later chapters of this book, elegantly established the fundamental nature of the new bodies that a cursory look might have revealed as mere curiosities. He not only saw mountains on the Moon, he recorded how their shadows changed as the Sun rose and set over the lunar landscape, and he measured their altitude as compared to terrestrial mountains. He not only discovered four moons of Jupiter, he patiently traced their orbits and determined their periods of revolution.

Galileo was driven by a faith in the ability of systematic observation to penetrate the innermost workings of nature. His eloquent, popular writings, written in Italian rather than the recondite Latin of medieval scholars, brought these findings to a wide audience, stimulating new ways of thinking about matter and motion, as well as about the order and structure of the Universe.

Looking through a telescope, indeed, was something anyone could do. *Seeing* through a telescope, however—and communicating that vision to others—was the mark of Galileo's true genius.

THE LEGACY OF GALILEO

After Galileo, as the telescope took its place as the quintessential tool of astronomy, clever optical designers began to look for ways to improve on Galileo's original design. The Galilean telescope, for instance, had a very small field of view, making it difficult to point accurately. Simply replacing the concave eyepiece with a convex lens, Johannes Kepler found, produced a telescope with a wider field. Its images were upside down, unlike those of Galileo's telescope, but that was not a critical drawback when looking into the sky, where no particular direction is more natural than any other. By the time Galileo died, the optical principles needed to design better instruments were well understood, and optical masters all over Europe were vying for excellence in the field. Telescopes became bigger in diameter, thus capturing more light, and longer, yielding higher magnification. Lens-grinding improved, resulting in sharper images.

One of the most important improvements in telescope design was made in 1704 by Isaac Newton. In place of lenses to collect and focus light, Newton recommended using curved mirrors. Mirrors had the advantage of being easier to make, since only one surface needed to be shaped, unlike a lens, which required two shaped sides. Mirrors could be made larger than lenses, since it was easier to cast metal than to produce disks of clear glass larger than a few inches in diameter. Most important, mirrors focused all colors of light at the same point, regardless of color. Lenses acted like prisms, bending light rays of different colors in different directions, thus making it impossible to avoid out-of-focus rainbows of light around brilliant objects like stars and planets.

Despite the clear advantages of Newton's reflecting telescope, lens-based telescopes, called *refractors*, continued to be used in astronomy in the 1700s and 1800s, because mirrors of those days were made of solid metal, an alloy called *speculum*. When speculum tarnished—which was very frequently—the entire mirror had to be taken out of the telescope and laboriously repolished. Nevertheless some patient astronomers were willing to put up with the trouble, building reflecting telescopes that collected more light than the smaller contemporary refractors. In 1781, William Herschel, using a reflecting telescope with a six-inch-diameter mirror of his own construction, discovered a new planet, Uranus, the first addition to the roster of the solar system since prehistoric times. Taking advantage of the great light-collecting power of his instrument, he went on to map the shape of the Milky Way by night-after-night counts of all the stars that passed through his eyepiece.

In the 1840s, William Parsons, the Earl of Rosse, constructed a telescope with a mirror more than 10 times larger than Herschel's: 6 feet in diameter. The monster telescope, erected at Lord Rosse's Birr Castle estate in Ireland, required a crew of two stout men and an elaborate system of pulleys and chains to point at objects, but it provided the first detailed views of the spiral structure of distant galaxies and the wispy clouds of gas in the Milky Way. It was so unwieldy, and the mirror so hard to maintain, that it soon fell into disuse, and, up to the first decade of the twentieth century, most astronomers continued to rely on smaller refractors for their research. The largest astronomical refractor ever built, commissioned in 1897, is the 40-inch-diameter telescope at Yerkes Observatory in Williams Bay, Wisconsin.

Until the middle of the 1800s, astronomers made observations simply by looking through the telescope and sketching or writing down what they saw. The invention of photography in the 1830s and 1840s made it possible to produce an instant and indelible record of observations, and by the end of the nineteenth century, observatories were beginning to compile libraries of images on the glass photographic plates that had become the staple tool of telescopic research. Photographic astronomy made it possible to discover moving objects with ease, leading to the discovery of great numbers of asteroids, small planetary objects that orbit the Sun, primarily between Mars and Jupiter.

Despite all these advances in instrumentation, astronomy remained in a relatively primitive state through most of the nineteenth century, largely because of the great distances that separate us from other objects in the solar system. Even through the telescope, most things look like small featureless points of light or faint, milky glows. The best astronomers could do was measure the positions of objects they saw and make a few sketches. They had observed a wide variety of objects, from asteroids to nebulae; they could predict eclipses and describe how planets moved. But they knew little more about what the stars or the moons of Jupiter were made of than Galileo did. In 1835, the French philosopher Auguste Comte lamented the impasse in astronomy: "In a word, our positive knowledge with respect to the stars is necessarily limited solely to geometric and mechanical phenomena."[4]

The doorway to a broader knowledge of the heavens opened at just about that time. Gustav Kirchhoff and Robert Bunsen, two German chemists who were investigating the colors produced when different chemical salts were ignited by a flame, invented the science we know today as spectroscopy. By breaking up light from a distant object with a prism and analyzing its distribution of brightness versus color (or wavelength), astronomers were able to deduce a wide variety of physical information about the source: temperature, chemical composition, velocity, rotation, and more. The old science of astronomy became the modern science of astrophysics, and astronomers began busily collecting information about the constitution of the planets, the composition of stars, and the motions of the galaxies.

The twentieth century saw the application of these techniques, coupled with remarkable leaps forward in telescope and other

---

[4] Cited in Laurence A. Marschall. *The Supernova Story.* Princeton: Princeton University Press, 1994. 17.

space exploration technology. Early in the century, astronomers solved the problem of building big reflectors. Instead of all-metal mirrors, they shaped their mirrors of glass, covering the glass with a thin, reflective film of silver or aluminum. When the metal became too corroded to use, perhaps once every few years or so, the old coating could be removed and a new one deposited, a process that could take as little as a few hours. Telescopes of 60, 100, and 200 inches in diameter were constructed, feeding light to the photographic cameras and spectroscopes of the new generation of astrophysicists.

During the first half of the twentieth century, using information from these new telescopes, along with physical insights gained from advances in atomic physics, nuclear physics, and optics, astronomers began to probe the structure and evolution of the stars, to measure the distances to galaxies, and to discover the age of the solar system and the Universe at large. Modern astronomy was about to enter a golden age.

The golden age of astronomy, when most of the advances reported in this book came about, was the last half of the twentieth century, especially the years after the launch of the first Earth satellite, Sputnik, in 1957. The pace of astronomical technology grew at an astounding rate during this period—so fast that we can only list a few of the major developments here.

One of the first developments was the expansion of the astronomical observing window to encompass the entire spectrum of electromagnetic radiation. Galileo's telescope collected visible light, but astronomers had learned over the intervening years that light was just the visible manifestation of electromagnetic radiation, a type of energy that includes radio waves, microwaves, infrared, ultraviolet, X-rays, and gamma rays as well. Objects in the heavens give off all types of this radiation, and by collecting these invisible glows from the heavens, we learn a great deal more about celestial objects than by just looking at visible light. Using radio telescopes after World War II, astronomers were able to map giant clouds of gas in the Milky Way and detect powerful sources, later shown to be black holes, in the centers of distant galaxies. Infrared telescopes penetrated through dark clouds of gas that were opaque to visible light, giving us our first glimpse of stars being formed.

Equally important was the ability to go into space. Space missions rapidly advanced our knowledge of the planets. By December 1972, we had a geologist on the Moon, and in the 1970s and the decades that followed, spacecraft were sent out to get close-up views of all the planets. As we write, there are second or later generation probes in orbit around Venus, Mars, and Saturn, another spacecraft visiting Mercury and still another on the way to Pluto. Self-contained robot explorers crawl across the deserts of Mars. Geologists now study not just terrestrial rocks, but pieces of rock from the Moon, and pore over pictures from the surfaces of Venus and of Saturn's moon Titan.

Space-borne telescopes, unaffected by the Earth's atmosphere, which blocks most forms of radiation from reaching the ground, have also afforded views of more distant objects. Ultraviolet, infrared, and X-ray telescopes report on conditions of bodies as nearby as the Sun and as distant as the quasars. Equally important, space telescopes produce images that are razor-sharp, unblurred by the Earth's atmosphere. The Hubble Space Telescope, launched by NASA in 1990, has given astronomers some of the most detailed and revealing images of the heavens ever achieved. Some have likened its effect to that of Galileo's first telescope, since both marked comparable factors of improvement in the detail astronomers could perceive.

Earthbound telescopes continue to develop, too, with new designs for mirror shapes that make them wider in their field of view, as well as more efficient in their use of light. New techniques of construction allow us to build mirrors that are made of dozens of sub-mirrors with which we can produce telescopes that collect tens of thousands of times more light than Galileo's did. We will discuss some of these new telescopes, and the prospects for future instruments, in later chapters of this book.

As a result of all these new tools, astronomers are now flooded with data and are able to collect more information in an eyeblink than Galileo could in three years of systematic observation at the eyepiece. None of these modern telescopes or spacecraft, needless to say, would function without electronic computers, the most important new tool of the astronomer. Computers run the telescopes, align the mirrors, collect the images, and analyze the data. They do what Galileo only dreamed of doing—they read the book of nature, written in the language of numbers.

It is a great new Universe we live in today. We began to discern its fuzzy outlines four centuries ago when Galileo's telescope first looked skyward, inaugurating a golden age of discovery and a revolution in science. Today, in another golden age of astronomy, the contours of the Universe have become far clearer, and many of its remarkable details are now known. We take this opportunity to look back at how it all began, where we stand today, and what the future holds in store.

CHAPTER 2

Telescopes

Two disks of glass and a piece of lead pipe: the first telescope was a simple device, but one that revealed things no one had dreamed of before. Telescopes were first used to amuse and amaze, but they changed the world, expanding the horizons of human consciousness far beyond the distances a rider could travel in a few days, to comprehend a universe measured in distances light traveled in billions of years.

Over 400 years, Galileo's simple design has gained immensely in both size and complexity. Astronomical telescopes today do the same things that the first telescopes did—make distant things look brighter and clearer—but the ones used by research astronomers today are giant machines moved by electric motors and controlled by computers. They concentrate the faint light from the heavens— just like Galileo's—but instead of focusing it through an eyepiece so that someone can look at it, modern telescopes turn the light into digital data.

Along with an increase in size and complexity has come an increase in the social enterprise involved in astronomy. Galileo built his first telescopes by himself in a week's time. A modern telescope is built by a large team of engineers, technicians, and scientists. Building it may take more than a decade from design to completion, and involves the casting and grinding of ultra-precise mirrors, the coding of software, and the assembly of giant supporting structures. Space-based telescopes are not complete until they are placed into orbit around the Earth or the Sun. Henry Ford's Model T doesn't look that much different from a modern automobile and was built on assembly lines that aren't much different in principle than those in operation today. But Galileo's 1609 telescope bears only a distant family resemblance to its counterpart of four centuries later.

GALILEO'S TELESCOPE

As we discussed in Chapter 1, Galileo Galilei did not set out to start a revolution in astronomy. In fact, prior to the autumn of 1609, when he first turned his newly constructed telescope skyward, the ambitious young Galileo was mostly occupied with down-to-earth pursuits. As professor of mathematics at the University of Padua, he wrote and lectured on a wide variety of technical subjects, many with a distinctly practical bent, such as the construction of military fortifications and the design of machines to pump water.

In the summer of 1609, travelers to Italy brought news that Dutch eyeglass makers had devised something altogether marvelous—a tube, which, when one looked through it, brought distant things closer. The details were rather vague, but Galileo quickly realized that such a far-seeing tube could be of enormous use, enabling merchants and princes to anticipate the arrival of ships returning to port and helping generals to identify the disposition and strength of the enemy long before battle was joined. In an age when scholars and artists alike depended on the financial support of wealthy patrons, Galileo was particularly eager to use his talent to curry favor among the rich and powerful. Devising a telescope—especially a more powerful one than that constructed by the Dutch—seemed an ideal project for this purpose.

Using what he knew about optics, Galileo was quickly able to cobble together something that did the trick. He attached two eyeglass lenses, one convex (curved outward) and one concave (curved inward), to opposite ends of a hollow cylinder made of lead. Looking through the optical tube made objects appear about three times larger than they did to the naked eye. By modern standards, the magnification was pretty feeble, and it took a little bit of practice to look straight through the tube at the proper angle to avoid seeing only its leaded walls, but Galileo knew that this was only the start. He'd proven the concept; now he had to get to work improving it. After tinkering for less than a week, he came up with a deluxe model, over twice as powerful as his first, which magnified objects by about eight times.

Eager to promote the wonders of his spyglass, Galileo took it to Venice, where he installed it at the top of the Campanile of St. Mark's. Curious Venetians who climbed the tower during this time could point the tube at tiny specks on the horizon, look through the eyepiece, and—to their great amazement—see sails and sailors as clearly as if they were lying at anchor in the harbor. They scanned the busy city, looking at church towers twenty miles away and spying on passengers disembarking from ferries. (To our knowledge, Galileo never got the idea of installing coin-operated models in busy tourist sites, but the appeal of the telescope to idle sightseers was immediately clear.)

After several days of demonstrating the telescope to crowds of onlookers, Galileo presented it as a gift to the Venetian Senate at the end of August 1609. The senators, duly impressed, promptly increased Galileo's salary and granted him lifetime tenure in his position. Thus, even before he used it for astronomical purposes, Galileo's telescope brought him fame and fortune.

Yet sometime in the autumn of 1609, perhaps merely by accident, Galileo must have pointed his telescope at something in the sky. It may have been the Moon, or the Milky Way, or the planet Jupiter. In any case, what he saw was so striking that he immediately began recording the remarkable things that the telescope revealed about the heavens. Those observations, which he first published in *Sidereus Nuncius*, were to change the prevailing worldview and open up the Universe to close scientific investigation.

Prior to this time, astronomers had no way to magnify or inten-sify the light from stars and planets—and the notion of bringing the heavens up close with lenses probably never occurred to them. With only their naked eyes to rely on, astronomers could not examine the surface of the Moon or planets, could not see objects if they were fainter than the limits of the human eye's sensitivity, and could not resolve the Milky Way into individual stars. Except for the Moon and an occasional comet, everything astronomers studied looked like a featureless dot of light. All they could do was to record the changing positions of the bright dots of the planets among the other dots of the stars.

The master of this type of astronomical observation in the pre-telescopic era was a Dane named Tycho Brahe. Under the patronage of the King of Denmark, Brahe built an observatory on the small island of Hveen not far from Copenhagen. There he installed unique astronomical instruments of his own design. Nothing but ruins remain now, but if you had seen the observatory during its heyday, you would not have recognized anything resembling the telescope tubes you see in modern astronomical domes. Brahe's instru-ments looked like large versions of the calipers and dividers used by machinists to lay out angles. For instance, several of his instru-ments sported two long brass arms, pivoted at one end, which could be adjusted to measure the angle between two stars. The observer would stand at the pivot end and sight along each arm until the stars in question lined up with a peg at the other ends of the arms. Then the angle between the arms could be read off a scale resembling a protractor that spanned the arc between them.

Brahe's angle-measuring instruments were monumental in size, dwarfing the observers who used them, and they cost a fortune to construct. Yet their large size was essential to their success: it made it possible to read the scales to the most precise division. Brahe's measurements of the positions of the planets, which he collected for three decades from his island observatory, were unexcelled in the era before the telescope—the best naked-eye observations ever made. Johannes Kepler used Brahe's records in the early 1600s to show that the orbits of the planets were not circles, as ancient astronomers had taught, but ellipses, and he could not have succeeded had not the measurements been so precise. Still, Brahe's measurements gave no information about things we consider part of astronomy today, such as the craters and valleys of Mars or the bands of color on Jupiter. He couldn't see these features at all—planets and stars alike were just bright pinpricks in the dark canopy of the sky. He couldn't even tell that the planets and stars were spherical, and neither could anyone else at the time.

Galileo's application of the telescope to astronomy was to change all that, though by modern standards, his telescopes were weak and tiny instruments. The 8-power telescope, which he presented to the Venetian Senate, had about the same magnifying power as an inexpensive pair of binoculars you can buy today and was considerably poorer in sharpness and ease of use. About two feet long and an inch and a half in diameter, and it was held in two hands, so that people using it looked like the standard image we have today of sailing ships' captains scanning the horizon with a tube to their eye.

As Galileo continued to tinker, his instruments improved in usefulness and power. The telescope he first used to observe the heavens several months after his trip to Venice magnified thirty times and was clamped to a stable mount to eliminate the unavoidable wavering of the original handheld telescopes. By the spring of 1610, when he published the results of his pioneering observations of the heavens, he had constructed over a hundred telescopes, most of which he sold or gave away, and he continued to make telescopes for much of his life.

Galileo's great contribution, clearly, was not to invent the telescope, but to improve it and to apply it to astronomy. However, a case can be made that he was one of the first—if not the first—to devise binoculars. His single-tube telescopes were fine, he found, for observing from fixed location, but sailors at sea had a hard time using them, because it was so difficult to hold them steady by hand. Even if the telescope was clamped to a stationary stand attached to the deck, the rolling of the ship in all but the calmest seas made it impossible to keep astronomical objects in view. So about 10 years after his first telescope, Galileo devised what he called the *celatone* or "large helmet," which must have been one of the strangest forms of the telescope ever devised. Two telescopes were attached to the front of a large headpiece. The user of the *celatone* put the headpiece over his head, which positioned the two telescopes directly in front of his eyes. Thus, by keeping his head pointed in the direction of the object he wanted to see, a sailor could compensate for the unsteadiness of the ship and keep the object in view.

With two large telescopes strapped onto his head, the user of the *celatone* must have looked like a large, ungainly insect. During tests, Galileo found that it was still difficult for an observer to compensate for the ship's roll, so he devised a chair that floated in a tank of water on the deck of the ship, a contraption that was supposed to provide a more stable platform for the observations. Needless to say, this didn't work too well. Galileo's strap-on binoculars never gained wide acceptance. In fact, probably no one other than Galileo himself ever used the *celatone*.

The telescope, on the other hand, gained immediate acceptance. Few of the pioneering telescopes that Galileo constructed survive, but two Galilean telescopes at the Museum for the History of Science in Florence are probably typical of what he built. The lenses are mounted at the ends of wooden tubes a couple of feet long covered with cloth and leather. Their proportions are more slender than most telescopes today, which means that images seen through them must have been relatively faint by modern standards. The larger the diameter of the telescope, the more light it collects, and the longer the telescope, the more it magnifies. Highly magnified images tend to be fainter, since the light collected by the telescope is spread out more, and thus the brightest images are produced by telescopes with large diameters and short lengths. But at the time of Galileo, it was difficult if not impossible to produce large-diameter lenses of clear glass. Though historians credit Galileo with extraordinary skill in fashioning lenses—making his telescopes better than most others of the time—they were still very small by modern standards.

Yet, as is the case with so many other inventions, the crude instruments that pioneered the field proved an inspiration to many other inventive minds. Many of us can recall when the first personal computers appeared in the 1980s, equipped with less than a megabyte of memory and giant black-and-white monitors, puny boat anchors compared with the laptops of today, which are thousands of times faster and more compact. So it was with the telescope. Over the 400 years since Galileo first turned his optic tube skyward, the design of the telescope has been improved, perfected, and transformed into a variety of specialized instruments. Without them we would be blind to the marvels of the vast universe that surrounds us.

TELESCOPES TODAY

As we touched upon in Chapter 1, today's most powerful research telescopes could not be more different from Galileo's. They gather and focus light with mirrors, not lenses. These telescopes are *not* long and skinny but short and fat in proportion, like searchlights. They are not intended (and often, not even equipped) for eyeball viewing, and so they have no eyepieces to magnify the images. They are pointed and steered by computers, and no one records the things they see in a sketchbook—they don't even record images on photographic film. The era of photography, at least for astronomers, ended in the early 1990s, and astronomical darkrooms no longer exist at most major observatories. The cameras and other instruments on current telescopes are equipped with electronic sensors, which turn incoming light into digital data that can be viewed on a computer screen as pictures or graphs.

And, of course, contemporary observatory telescopes dwarf Galileo's perspicillum. The 300-ton Keck Telescope, 8 stories tall, sits atop Mauna Kea on the Big Island of Hawaii at an altitude of 4,200 meters (13,796 feet). It has a main mirror 10 meters (33 feet) in diameter that collects 69,000 times more light than Galileo's telescope. The Keck also produces much sharper images, thanks to its much larger optics, its high-altitude location, and a system that applies carefully calculated forces twice per second at many positions on the back of the main mirror (actually composed of 36 separate 72-inch segments). These forces bend the upper (reflecting) surface of the mirror to adjust for shifts in position and shape of the mirror segments, which sag as the telescope points at different angles with respect to the vertical direction. In addition, a laser projects a spot of light on a layer of sodium atoms 55 miles above the Earth, providing a reference beacon for an electro-optical system. As atmospheric effects blur the known shape of the laser spot, the system bends a small flexible mirror in the telescope to compensate for the blurring.

And less than 300 feet from the Keck Telescope (actually, the Keck I Telescope), there is another just as big, the Keck II.

Elsewhere on Mauna Kea, there is the 8-meter (26-foot) Gemini North telescope, the 8.2-meter (27-foot) Subaru Telescope that belongs to Japan, the smaller Canada-France-Hawaii Telescope, and many more. In Chile, the European Southern Observatory (with headquarters across the ocean in Munich) operates two large mountain observatories, one with its star attraction, the Very Large Telescope (VLT). The VLT comprises four independently mounted 8-meter telescopes, and the member nations of the ESO include Italy; thus, Galileo's direct successors in astronomy can use telescopes perhaps beyond his imagination.

To astronomers like the authors, who grew up in a period when the most powerful telescope on Earth was the post–World War II 5-meter (16-foot) telescope on Palomar Mountain, the pace of innovation and inflation in telescope size seems almost as unimaginable as the 16-foot telescope might have seemed to Galileo. Now on the drawing boards or even under development are behemoths such as the Giant Magellan Telescope (GMT), the Thirty-Meter Telescope (TMT), and the European Extremely Large Telescope (E-ELT).

The GMT will consist of seven large mirrors on a common mount, with a combined collecting area equal to that of a single 24.5-meter (80-foot) telescope. It may be in operation by 2016 and yield images that designers claim will be up to 10 times sharper than those from the Hubble Space Telescope, yet achieved from the ground. A rival project, the Thirty-Meter Telescope would be even bigger, composed of 700 smaller mirrors that together equal a 30-meter (98-foot) light-collecting surface. The U.S. government is likely to choose one of these ambitious efforts for federal funding and then change the name.

European astronomers may exceed whatever the United States does in future giant telescopes, unless we join hands with them. They were granted 57 million euros in December 2006 to begin design studies on a European Extremely Large Telescope of about 40 meters (131 feet) that could be under construction by 2010. (It won't be in Europe; there's no observatory site of sufficient quality on the continent.)

Galileo worked by pointing his telescope at a recognizable object that was already visible to the naked eye, such as Jupiter. In the lingo of modern photography, he had a "point-and-shoot" device, but there was no photography yet, so it was just "point-and-look." He sketched and wrote down what he saw. If the telescope revealed an interesting faint object alongside Jupiter, Galileo could watch only for a brief time, until Jupiter's motion across the sky took it away from the interesting body. He had no way to locate an object like that when it was no longer near the line of sight to Jupiter or another bright object. (His notebooks show that he came upon Neptune this way, over two centuries before it was officially "discovered.")

Modern computerized telescope controllers direct the telescope at the known sky coordinates of a celestial body or at predetermined positions in a search pattern. They track the motion of that target across the sky as the Earth turns. When observing an asteroid, comet, or other object that moves across the background pattern of the constellations, the telescope automatically tracks that interplanetary object according to its known orbit.

Galileo used his telescope in town. Today's major telescopes are on mountaintops or up in space for clearer views against darker skies. Astronomers cannot and do not need to look through the big telescopes, and therefore, they are often detached from them. The observer using a mountain telescope might be on a lower floor of the observatory building or at a convenient field station thousands of feet below the summit. Often, he or she is dozens, hundreds, or thousands of miles away, watching a computer screen that displays the images and other readings from the telescope. In some cases, observers are not even watching the process. Instead, they have specified the desired observations in a "script" that lists the target star, the instrumental settings to be used, the desired exposure time, and more. In due course, when the observations are accomplished, a data tape or disk, an e-mail, or perhaps a file posted on a secure Web site will present them with the image and other data that the telescope gathered. That's how it works, of course, when they use an orbiting instrument, like the Hubble Space Telescope. There are rare cases in which the observer is stationed in the control room, but most of the time, he or she is "offline" at his or her home university, attending a conference, or maybe even sleeping or catching a flick at the cinema. He or she will get the results of the observations in due course. The rare but wondrous eureka moment when an observation yielded an almost inconceivable discovery that sometimes came at the telescope or at least on a viewing monitor hooked up to the telescope is now almost a thing of the past.

As we approach the 400th anniversary of the astronomical telescope, many astronomers are deserting the very concept of using a telescope, robotically or otherwise, to make their own uniquely planned observations. Instead, an increasing number of scientists *don't* script a robotically controlled observation as we just described. (That would be "so '90s.") The new fashion is to rely on observations that are made without regard to specific needs by tireless, computerized telescopes that survey the heavens in a systematic manner. These automated survey telescopes scan the sky throughout the night for weeks, months, and years. The observations are entered into huge digital databases that can be accessed remotely from an astronomer's desk. These astronomers, called *data miners*, organize specialized computerized searches that sift through a database, seeking small nuggets of new knowledge in vast stores of celestial records. Some data miners, for example, seek information on obscure dwarf galaxies among bigger and brighter galaxies. Others prospect for extremely old stars lurking among hundreds of millions of relative stellar youngsters in the Milky Way. Still others dig through the databases in search of icy dwarf planets beyond Pluto on the outskirts of the solar system. In each case, they investigate unknown aspects of the heavens, some already awaiting detection in accumulated sky survey data, but not yet recognized. In the spirit of the TV quiz show *Family Feud*, they program a computer to determine what the "survey says."

The Sloan Digital Sky Survey, with a 2.5-meter (8-foot) telescope at Apache Point Observatory high in the Sacramento Mountains above Alamogordo, New Mexico, is billed as "the most ambitious astronomical survey ever undertaken." The survey is intended to produce a three-dimensional map of about one-quarter of the whole sky visible from Earth. By June 2005, it had photographed over 200 million celestial objects, including stars and more than a half-million galaxies, each consisting hundreds of billions of stars. In fact, current estimates for the number of stars in our own Milky Way galaxy are in the low trillions. In 2003, data miners delving into the Sloan data found independent evidence for the previously reported "dark energy," a mysterious force that is making the Universe expand faster and faster. A *New York Times* account of these findings declared "Astronomers Report Evidence of 'Dark Energy' Splitting the Universe." This alarmed the *Late Show* TV host, David Letterman, who summoned a cosmologist from Carnegie Mellon University to explain the findings. Letterman didn't understand dark energy, and he professed astonishment that the *Times* would bury on page 13 the fearsome news that the Universe is being ripped apart.

Hundreds of other studies have come from data mining in the Sloan Digital Sky Survey, which is being continued and expanded with improved instrumentation, but already its prominent status has been challenged. According to Professor John Tonry, an astronomer at the University of Hawaii, by mid-2007 projects like the once-unprecedented Sloan Survey could be regarded as "limited in temporal or angular coverage," meaning that they did not monitor a given part of the sky frequently enough and did not observe a sufficiently large region of the sky according to standards that were set by subsequent celestial surveyors.[5] However, "continuous observations of the entire sky will soon be routine" with new survey telescopes,

---

[5] Tonry, John. "Synoptic Sky Surveys and the Future of Astronomy," invited talk at the 210th meeting of the American Astronomical Society, Honolulu, Hawaii, May 28, 2007.

according to Tonry's opening address to a meeting of the American Astronomical Society in Honolulu in May 2007.

A new survey, called the Panoramic Survey Telescope and Rapid Response System (Pan-STARRS), will use four fairly modest telescopes, just 1.8 meters (6 feet) in diameter, each equipped with one of the world's largest digital cameras, rated at 1.4 billion pixels. The huge amount of data that Pan-STARRS will begin generating at its chosen location atop Haleakala Mountain on Maui in 2010 will serve many different astrophysical investigations, but it is likely to attract the most attention for its declared intention to search for previously unknown "killer asteroids" that might someday strike the Earth.

Perhaps the most extreme difference between Galileo's telescope and those of the current era is presented by the many kinds of telescopes that operate outside the range of visible light. The great 1,000-foot bowl-shaped radio telescope at Arecibo, Puerto Rico, is one of the most famous. The big dish is equipped with radio receivers that observe the cosmos at frequencies from about 300 MHz to 10,000 MHz, and it is capable of bouncing radar pulses off planets, moons, and even the Sun. Another great radio telescope, the Very Large Array (VLA), is a system of 27 separately mounted dish antennas, each 25 meters (82 feet) in diameter. The VLA is on the Plains of San Agustin in central New Mexico and is a neat place to visit. In the science fiction film *Contact*, a radio astronomer played by the actress Jodie Foster used both the VLA and the Arecibo telescope to search for extraterrestrial intelligence. Those were fictional efforts, but real projects in the Search for Extraterrestrial Intelligence are underway at several radio telescopes.

The Hubble Space Telescope is only 2.4 meters (8 feet) in diameter, but by virtue of its superb viewing location hundreds of miles above the surface of the Earth, it is capable of seeing farther and fainter objects than any previous telescope. It was equipped to observe the heavens in visible and ultraviolet light, and, after a new instrument was installed by Space Shuttle astronauts in 1997, it operated in infrared light as well. A companion Spitzer Space Telescope was launched in 2003 and is trailing the Earth in its orbit around the Sun making infrared observations.

Other orbiting telescopes, and those launched into space beyond Earth orbit, explore the cosmos in still other forms of light, including X-rays, gamma rays, and submillimeter arrays. Each form of light is typically generated by different kinds of objects in space, or by matter in different physical states within the same object. The Sun is a simple example: astronomers who want to view sunspots find that they are best seen in visible light. Explosions on the Sun show up most strongly in X-rays, ultraviolet light, and in extreme cases, gamma rays. Typically, the more energetic the form of light (the shorter the wavelength of the radiation), the hotter the gas that produces it.

The behemoth among space telescopes is the James Webb Space Telescope (JWST). With a huge folded mirror (that will hopefully unfold and assume exactly the right position after launch) and a sunshade the size of a tennis court, it will search for the earliest stars and galaxies in the history of the Universe from a viewing position about a million miles from Earth in the direction opposite the Sun. The shade will keep the heat of the Sun and the Earth off the sensitive telescope and instruments, which must operate at a temperature of -370°F (50 kelvin) to detect faint radiation from the far Universe. Antennas mounted on the spacecraft "bus," on the Earthward side of the sunshade, will receive commands from ground controllers and send back a stream of digital data from the telescope. By the expected time of launch in 2013, the Hubble may have ceased operations, and the JWST will be the flagship telescope in space.

Suppose one could transport Galileo from his workshop in 1609 to the giant telescopes at work on mountaintops today. He might not be willing to go back.

CHAPTER 3

The Moon

THE MOON IN THE SEVENTEENTH CENTURY

Who does not know the face of the Moon? Even in Galileo's time people untutored in astronomy regarded it as the most familiar sight in the nighttime sky, changing shape and brilliance regularly as it went through its monthly cycle of phases. Learned men had devised methods for estimating its size and distance—about 2,000 miles in diameter and about 240,000 miles from Earth—that were not far from the modern values. Academics had long speculated on the nature and composition of its surface. It was natural, therefore, that the Moon would be the first heavenly target of Galileo's telescope.

Surprisingly, however, Galileo was in no hurry to turn his telescope to the nighttime sky. Terrestrial uses for the telescope initially seemed far more promising to him; the ability to spy on distant objects (like armies and warships) had obvious military and maritime applications. The delighted crowds who looked through Galileo's first telescope from the tower of St. Mark's in August 1609 gave no thoughts to using it as an aid to astronomy; nor did Galileo. He knew that the commercial success of his telescope depended on its power and ease of use to merchants and military men. Accordingly, he devoted himself to improving the magnification and clarity of the instrument, eager to establish himself as the reigning expert in the new technology.

Throughout the fall, therefore, Galileo was occupied with grinding lenses, and if he took time to look through his telescopes, he found plenty of targets in distant buildings and unsuspecting citizens. It was not until the beginning of December 1609 that Galileo began to point a telescope skyward, and, virtually overnight, the course of his life was changed: Galileo the military engineer became Galileo the astronomer.

The Moon was probably the first object that Galileo looked at in the sky, and, despite all he knew about the Moon—or thought he knew—the erstwhile professor could not restrain his amazement. If his eyes were to be trusted, there were mountains, valleys, and plains covering the entire surface of the globe. The Moon, in short, had a landscape.

That was most peculiar. Galileo had learned from his schoolmasters what every educated person in the sixteenth century took as well-established fact—that the Moon was a very different sort of place from the Earth, distinguished by its very lack of distinguishing features. The lunar surface, according to the common wisdom, was supposed to be as smooth as the shaven head of a monk. Belief in an unblemished Moon had its origins in the science of Aristotle, whose worldview had guided the educational establishment for the previous 1,800 years. Aristotle taught that bodies in the heavens had absolutely nothing in common with the Earth, either in substance or in appearance. Heavenly bodies were perfect, while terrestrial bodies were rough, irregular, and corrupt.

It isn't hard to see what Aristotle was getting at. On Earth, growth and decay are the norm. Things are always changing; we can never step twice into the same river. Volcanoes erupt, mountains are worn down by the weather, flowers bloom and wither, animals are born and die. The only thing that does not change, perhaps, is change itself.

The Earth is changeable, in the Aristotelian view, because everything on Earth is made of four basic elements—earth, air, fire, and water—mixed together like ingredients in a shaken bottle of salad dressing. Left to interact, they spontaneously rearrange themselves according to their natural tendencies: earth and water tend to sink toward the center of the Earth, and air and fire tend to float upward. The burning of a candle, the sprouting of a seed when watered, and the crumbling of a rock are examples of the four elements shaking themselves out of the mixture they find themselves in and seeking their natural resting places in the cosmos.

But heavenly bodies—and the Moon was the closest of the heavenly bodies—were thought to be entirely composed of a fifth substance not found on the Earth. This special stuff—the *aether* or *quintessence*—was perfect, eternal, and unchangeable. Bodies made of aether traveled in perfect circles around the center of the stationary Earth, because things moving in circles repeated the same path over and over, never changing their motion. Anything made of aether would, it follows, be perfect in appearance as well—a perfect sphere in the case of the Moon and the planets.

Perfection meant that heavenly bodies were never subject to change, to cracking or bulging, or to wearing down by wind and water. It also meant that heavenly bodies were completely symmetric—perfect spheres, uniform in every way, with nothing on their surfaces to distinguish one part from another. The Moon, of course, didn't appear quite uniform, even in the days before the invention of the telescope. Since ancient times, people had noticed irregular dark areas on the face of the full Moon, which some called "the man in the Moon" or the "face on the Moon." (Chinese skywatchers seeing two markings that suggested long, floppy ears called the pattern a "rabbit in the Moon.") In the 1590s, two decades before the introduction of the telescope, Englishman William Gilbert prepared a map of the dark and light patterns he saw on the Moon with his naked eye, and proposed that the light areas were water and the dark spots land.

Gilbert's suggestion that the Moon had seas was not new, though most previous writers had identified that the dark areas, not the light ones, as bodies of water. Even today, though we know the Moon has no bodies of water at all, these dark areas are called *maria*, from the Latin word for "seas."

Scholastic astronomers in the Middle Ages argued quite a bit about whether the man-in-the-Moon markings—whatever they were—contradicted Aristotle's notion of heavenly perfection. Some suggested that they were not on the Moon at all but were caused by clouds on the Earth. Others maintained that they were shadows cast by uneven amounts of light from the Sun. And still others were willing to attribute the patterns to a slight variation in the reflective power of the lunar surface from place to place—a variation that somehow was not a serious violation of the perfect uniformity of heavenly bodies. But whatever the explanation, the Moon was still supposed to be a flawless example of the perfection of the heavens.

And so, turning his telescope on the Moon in December 1609, Galileo was startled to behold a world that looked uncannily like the Earth. In a letter he wrote to a friend in January 1610, following a month of intense lunar observations, Galileo marveled, ". . . it is seen that the Moon is most evidently not all of an even, smooth, and regular surface, as a great many believe of it and other heavenly bodies, but on the contrary it is rough and unequal. . . . it is full of prominences and cavities similar, but much larger, to the mountains and valleys spread over the Earth's surface."

Galileo didn't take what he saw at face value, and he realized that extraordinary claims require extraordinary evidence. The nature of the landscape he saw amazed him so much that he felt he had to prove, even to himself, that he wasn't somehow deluded or hallucinating. He observed the Moon repeatedly and patiently for at least a month, spending hours watching and sketching, as is clear from the description he gives in *Sidereus Nuncius*.

According to Galileo's account, the first thing that struck him was that the terminator, the line dividing the daytime and nighttime regions of the Moon, was not smooth, as would be the case if the Moon were a perfect sphere, but was jagged and uneven. There were dark bays that extended out into the lighted face of the Moon, just like shadowed valleys seen from the top of a mountain in the late afternoon. On the nighttime side of the terminator, he could see spots of light, speckling the darkness like fireflies. As the phases of the Moon changed, the terminator moved across the face of the Moon, and the spots of light that lay on the dark side of it grew and eventually blended into the advancing daylight. This was just what would be expected if the spots of light were the tops of mountain peaks, bathed in light by the rising of the Sun: "Now on Earth, before sunrise, aren't the peaks of the highest mountains illuminated by the Sun's rays, while shadows still cover the plain?"

By applying some simple geometry, Galileo was even able to estimate the altitude of the sunlit tops of the mountains. All he need to know was the distance that separated the sunlit peaks from the terminator—the higher the mountain, the more it would stick out from the surface of the Moon, and the more distant it could be from the terminator and still catch rays of the Sun around the curve of lunar globe. Galileo was surprised to find that highest lunar mountains were about four miles in altitude, taller than the tallest known mountains on Earth, which were then thought to be only a mile or so in height. (In Galileo's time, geology, like astronomy, was still in a primitive state, and explorers had yet to measure, much less scale, the four- and five-mile-high peaks of the Andes and Himalayas.)

Galileo also was the first astronomer to see and describe craters, which speckled the surface of the Moon, commenting that they decorated its surface like the spots on a peacock's tail. They were abundant everywhere, though less frequent in the slightly darker areas that make up the shape of the man in the Moon. The darker areas thus appeared to be flatter, smoother plains, while the rest of the Moon was more rugged and irregular.

Craters, which seem so characteristic of the Moon to the modern mind, were a puzzle to Galileo and others who saw them for the first time. Were they in fact just spots, or were they parts of the relief of the landscape? Again the pattern of light and shadow provided an answer: Galileo noted that craters near the terminator always appeared to have one side in shadow, and the shadowed side was always the one closest to the lit side of the terminator. This suggested that the craters were more than just circles on the ground, but real depressions, surrounded by rings of mountains that cast their shadows on the crater floors.

Galileo's telescope helped clarify one other mystery of the Moon, a puzzle of long standing. At crescent phases, when most of the visible surface of the Moon lay in darkness, a faint glow could be seen emanating from the darkened side. The smaller the crescent, the brighter the glow appeared, so bright in fact that through his telescope Galileo could see the outlines of the craters he had sketched a few weeks earlier when that part of the Moon's surface was in full sunlight. In the years before the telescope, other observers had noted this ashy light, which some called "the old Moon in the new Moon's arms." They supposed that it was a sort of luminous glow produced by the Moon itself. Not so, remarked Galileo, for the ashy glow was not visible during lunar eclipses, when direct sunlight on the Moon was blocked by the Earth's shadow. Now that he could see it clearly through the telescope, it seemed natural to conclude that the glow came from above. If you were standing on the dimly lit "old Moon" portion of the crescent Moon, there would be something shining down on you from the sky.

The only body that could cause the glow, Galileo reasoned, was the Earth, for when the Moon is near its crescent phases as seen from the Earth, the Earth, as seen from the Moon, is nearly fully lit and therefore lights up the lunar surface. "In an equal and grateful exchange," he wrote with a subtle touch of humor, "the Earth pays back the Moon with light equal to that which she receives from the Moon almost all the time in the deepest darkness of night."[6] Galileo was right, and today we call this phenomenon of sunlight reflected back onto the dark part of the Moon by an appropriate name: "earthshine."

Even then, there were many who disputed these observations that seemed so conclusive to Galileo. Too much was at stake— the perfection of the heavens had been central to the teaching of astronomy for over 1,500 years. One astronomer, J.G. Brengger, read

---

[6] Galilei, Galileo. *Sidereus Nuncius or the Sidereal Messenger.* Albert van Helden, trans, Chicago: The University of Chicago Press, 1989. 55.

the description of the Moon in *Sidereus Nuncius* with alarm. Brengger was willing to grant that the Moon *appeared* to have a rough and earthlike surface. Nevertheless—rest easy, Aristotle!—it really wasn't like that. Let us suppose, he argued, that the Moon were covered with a crystal substance, perfectly transparent, that filled in all the valleys and craters out beyond the height of the highest mountain on the Moon, and that the outermost surface of the crystal were a perfectly smooth sphere, holding the apparently irregular Moon like an insect embedded in a globe of amber. That way the Moon could still be perfectly smooth—as Aristotle required it to be—and still seem to be earthlike.

Galileo, informed of Brengger's proposal, fulminated and scoffed, pointing out that this was a logic of desperation: how could anyone invoke an invisible crystal sphere to draw any conclusions about the shape of the Moon? The Moon was a tangible world, looking more and more like a smaller version of the Earth under the magnifying eye of the telescopic lens. Study the Moon with telescopes, Galileo seemed to be saying, and leave aside all this idle speculation about invisible crystal spheres!

To be sure, Galileo's own study of the Moon was, by modern standards, rather crude and perfunctory. He made only a handful of sketches of the Moon, and though they show the overall irregularity of the lunar surface, there are very few markings that can be matched exactly with modern maps. A large crater straddling the terminator in several of his drawings is clearly much bigger than any actual landmark on the Moon. Galileo apparently exaggerated its size to emphasize, through the pattern of shadow and sunlight on the crater floor, that it is a significant depression in the lunar surface. But he never took the time to plot a careful map of the new world his telescope had revealed.

Inspired by Galileo, other patient observers, equipped with improved versions of Galileo's instrument, rose to the challenge of lunar mapping. The first telescopic lunar chart, published by the Dutch cartographer Michael Florentius van Langren in 1645, assigned names to 270 lunar maria and craters. Two years later, the Polish astronomer Johannes Hevelius published a monumental study of the Moon, called the *Selenographia*, in which he first assigned the names of terrestrial mountain ranges, like the Alps, Apennines, and Pyrenees, to various lunar features. Each of the early lunar mapmakers devised his own system of naming places on the lunar surface, sometimes honoring terrestrial cities and landmarks, and sometimes honoring patrons, kings, and fellow scientists. While some of these early names—like those of Hevelius's mountains—are still used today, it was not until the twentieth century that an international system of lunar nomenclature could be agreed upon. Still, the propensity of the new generation of lunar observers for naming features on the Moon after terrestrial people and places was indicative of a watershed in scientific thinking. No longer was the Moon considered a different type of place in a heaven quite distinct from the Earth. Both the Moon and the Earth were made of the same sort of stuff. Both were worlds circling in a universe whose wonders were only beginning to be explored.

Paradoxically, while Galileo's telescope opened up a new era of lunar observation and mapping, it did not solve many fundamental questions about the Moon. Why were some portions—the maria—darker than others? What were the causes of the craters? Was there any water on the Moon? Were there creatures inhabiting it? As late as the mid-nineteenth century, reputable scientists debated the habitability of the Moon, and it was not until the manned and unmanned Moon landings of the 1960s that a firm understanding of the Moon's history and geology finally began to take shape.

On that December evening in 1609, when he first pointed his telescope at the familiar glowing disk of the Moon, Galileo joined the ranks of the great explorers of the Age of Discovery: his new world, the Moon, lay not across 4,000 miles of ocean, but across a quarter million miles of space.

THE MOON TODAY

The Moon as we know it now was unimaginable in Galileo's time. It is bombarded by rocks from space, sprayed with subatomic particles, and subject to huge swings in temperature. The stark lunar landscape is pockmarked with impact craters from collisions with large meteoroids and asteroids. Asteroids had not even been imagined in the seventeenth century, and it was not known to Western scientists that rocks fall from space. (Meteoroids are rocks traveling through interplanetary space. When they land somewhere, they are called meteorites.) Even at the microscopic level, the Moon is perhaps beyond the imagination of Galileo and his fellows. There are tiny pits in the rock, the little craters produced by the smallest and most numerous meteoroids falling from space, so-called micrometeoroids. Meteoroids, even these tiny ones, collide with the Moon at speeds that would make Superman envious—not just faster than a speeding bullet but hundreds of times faster.

The large dark areas that we see in the "man in the Moon" were long ago named as seas, like Mare Tranquillitatis (Latin for "Sea of Tranquility"). (That's where the first men on the Moon, Neil Armstrong and Buzz Aldrin, landed, dubbing their location "Tranquility Base.") They were called seas because they are dark and relatively smooth as seen through a small telescope. But they are dry. They are, in fact, volcanic plains. But they were not made by lava spilling from volcanoes as happens on Earth. They are the result of molten rock that welled up from the depths of the Moon 3 to 4 billion years ago. The lava emerged through parts of the Moon's crust that were weakened much earlier by huge impacts, so it came up through the floors of the largest craters, or "impact basins," and flooded the basins. Then it cooled and hardened.

The mountains of the Moon that Galileo discovered are not like those on Earth. They are just the jagged rims of craters, debris thrown up by the impacts that made the craters. There are no peaks on the Moon that were pushed up by the collision of tectonic plates impelled by continental drift, like the Andes or the Himalayas. Nor does the Moon have volcanoes like Mount St. Helens or Mount Etna, although it does have low, rounded hills that probably formed by volcanic processes.

The surface of the Moon is covered with fine mineral dust, or "lunar soil," but it is not like the dirt in your yard. It contains no insects, microorganisms, or organic matter that could nourish life.

Occasionally the dust rises up, reaching heights that may exceed a mile, but it is not like a sandstorm in the Sahara, where wind bears the particles aloft. Apparently the Moon dust is levitated by electrical force.

Moon dust sticks to everything. Much of it consists of sharp, angular rock particles that are hazardous to people, machines, and delicate surfaces, like those of telescope lenses. It will stick to solar panels and reduce the amount of electricity that they generate from sunlight. If a fair dose of Moon dust got into your lungs, your eyes, or your stomach, it would do you ill.

Atop a thick crust of relatively light (low-density) rock, the lunar surface is mostly regolith, a jumbled mixture of small and large rocks and rock particles. In typical land areas on Earth, most rocks come from nearby geological formations, although some may have been carried to their current locations by Ice Age glaciers or tumbled downstream in a swift-flowing river. But at any location on the Moon, the regolith is a jumbled mixture of rocks from all over. Impacts send rocks flying from one part of the Moon to another, seeding the regolith with debris from distant craters and with fragments of meteorites.

There is no sedimentary rock on the Moon like the layers of sandstone and limestone in the Grand Canyon. There is no sea, river, or lake on the Moon, and there never has been, so there's no water-deposited sediment. Many of the Moon rocks are breccia, which are conglomerations of fragments of many different earlier rocks that were broken apart and then welded together under the force of repeated impacts.

The weather on the Moon would have interested Galileo, if he could have measured it. There's no wind, the barometric pressure is virtually zero, the sky is cloud free (other than occasional levitations of Moon dust), and there is no rain. But if you are on the Moon, the temperature commands your attention. The heat comes from the Sun and is life threatening. Sunlight makes the temperature rise above 242°F during the day—it varies a bit from time to time and place to place—and there's never a cooling breeze. At night, there's no greenhouse effect to hold in heat that the Moon absorbed during the day, so the thermometer plunges to -270°F or even less. Conditions on the Moon make Death Valley and the Antarctic ice sheet seem downright hospitable.

At least there is no humidity on the Moon. Even on the hottest day, you wouldn't say, "It's not the heat; it's the humidity," as we often do on Earth. On the Moon, it is the heat; it's *never* the humidity.

The Sun's ultraviolet rays are another lunar hazard. There's no ozone layer on the Moon to shield it from the solar ultraviolet, as there is on Earth.

Other forms of radiation bathe the Moon as well. There are cosmic rays—subatomic particles from across the Milky Way. They strike the Moon everywhere and all the time. There are high-energy protons from occasional explosions on the Sun, which rain down on the Moon and can impart a fatal radiation dose to an astronaut. And there is the solar wind, a gentler outflow of lower-energy electrons and protons and other elemental particles from the atmosphere of the Sun. The solar wind strikes the lunar surface during two-thirds of every month. During the rest of the month, the Moon is passing through the outermost part of the Earth's magnetic field, which deflects the solar wind. When they do reach the Moon, some solar wind particles embed themselves in the lunar soil.

A lunar day is much longer than our 24-hour Earth day. It's about 27 Earth days, or roughly the time that it takes the Moon to make one orbit around the Earth. So when the Sun rises and it gets hot on the Moon, it will be almost two weeks before sunset and an abrupt drop in temperature.

There are a few places on the Moon where the Sun never shines. They are in the bottoms of craters near the north and south poles of the Moon. The Sun is so low in the sky at the poles (just as it is on Earth, but lower) that some parts of some crater bottoms are always in shadow during the day. They are in perpetual darkness, and much colder than anywhere else on the Moon, at about -370°F.

Some researchers think that, by remote sensing, they have detected ice on the Moon in these frigid craters, but the findings are inconclusive. If there were exposed ice, it would evaporate (technically, it would sublime, turning from a frozen state to a vapor, as dry ice, or frozen carbon dioxide, does on Earth). A sufficient layer of Moon dust and regolith might insulate ice in the crater bottoms, however, so it lasted for eons. Comets are made largely of ice, and they strike the Moon now and then just as meteoroids do. Perhaps the ice near the lunar poles (if any) came there aboard comets. NASA intends to look for it.

They're not what we would call "weather," but there seem to be occasional puffs of gas coming out from below the surface of the Moon. They may be a faint consequence of what little geophysical activity occurs inside the Moon. Compared to Earth, which has ongoing continental drift that causes mountains to rise, major earthquakes (like the magnitude 8 quake in Peru in August 2007), active volcanoes, serious beach erosion, and much more, the Moon is almost a dead world, with only these puffs and weak tremors, or "moonquakes."

There are tides in the solid body of the Moon, caused by the gravity of Earth, just as the ocean tides on Earth are largely caused by the Moon (the Sun causes weaker ocean tides). You wouldn't notice the tides if you were standing on the Moon, but they are readily detected by sensitive instruments. Ocean tides become stronger at certain times during the month, and so do tides in the Moon. Seismometers installed on the Moon by NASA astronauts detected moonquakes that recur regularly every month, triggered by the tidal force of Earth. Other moonquakes occur when meteoroids strike the Moon. The collisions make it shake. The seismometer readings reveal that on such occasions, the Moon rings like a bell, although you can't hear it. The total energy of seismic events on the Moon is probably less than one-trillionth of that on Earth. The Moon is a hostile world, but at least moonquakes are much less dangerous than earthquakes.

In addition to the maria of the Moon, there are bright regions, called the lunar highlands. The highlands are brighter than the maria because they are not paved with hardened lava, which is darker than other Moon rock. The highlands are heavily cratered but have no huge impact basins. There are more craters per square hundred miles in the highlands than in the maria, and many craters overlap each other because of impacting bodies that fell at adjacent locations. At one time, all or most of the Moon probably looked like the highlands. Then several truly huge impacts created the maria, eliminating all the preexisting craters within these impact basins. Of course, meteorites kept hitting the Moon, so newer, smaller craters were created within the maria. When lava later flooded the basins, it wiped out the small craters within. Then, over time, more impacts made new craters inside the basins, but the highlands have more craters because none have been wiped out by flooding. Thus, the relative ages of lunar regions are estimated by counting their crater densities (the number of craters of a given size per hundred square miles, for example). The more craters there are, the older the region. Newer terrain has fewer craters, because it was resurfaced more recently by a big impact or lava flooding. Small craters within a larger one are newer than the big one, though sometimes a ghost crater, or the outline of an older crater that hasn't been completely obliterated, survives.

Much of what we know about the Moon comes from laboratory studies of 842 pounds of Moon rocks and lunar soil that were returned to Earth by six manned expeditions of NASA's Apollo program, which visited the Moon from 1969 to 1972. We also gained valuable information from three robotic sample-return spacecraft, the Luna missions, launched by the Soviet Union. (The total weight of the three Luna samples is very small, but the samples have considerable scientific value.) Most of our other information comes from instruments planted on the Moon by Apollo astronauts, including seismometers and solar wind collectors and sensors, and from studies by a slew of robotic spacecraft that landed on the Moon, orbited it, or just flew by it.

The lab studies of Moon rocks revealed how old they are, what chemical elements and minerals compose them, and what the physical conditions were like where and when the rocks formed. (Radioactive dating tells when minerals crystallized from a molten state.) The Moon rocks range in age from about 3 to 4.5 billion years. Even more information comes from a few dozen meteorites that (unlike most meteorites, which come from the asteroid belt beyond the orbit of Mars) are rocks that were knocked off of the Moon by powerful impacts and later fell to the Earth.

All measurements, observations, and laboratory analyses lead to an unexpected conclusion: the whole lunar surface, down to a depth of many miles, was once molten. The original surface of the Moon with its first chemical composition was lost in the magma ocean. In this "ocean" state, lighter rock rose to the surface, cooled, and formed the crust of the Moon about 4.35 billion years ago. Then the features that we now see on the Moon began to form. Many of them were created in an era of very heavy asteroid and meteoroid bombardment that lasted until several hundred million years after the magma ocean solidified. The last few very big impacts created the present impact basins we have described.

Where did the Moon come from? Some astronomers thought that it was spun off from the newborn, supposedly molten and rapidly rotating Earth. Another idea was that the Moon formed from the birth cloud of the solar system in the same way that the Earth did, right alongside the Earth. Still another theory envisioned that the Moon formed elsewhere in interplanetary space, was captured long ago by the Earth's gravity, and has orbited Earth ever since. Each theory has implications for the composition of the Moon, compared to that of the Earth and other planetary bodies, and for the physical state of the Earth–Moon system (for example, the speed at which they are rotating). Each theory has suffered fatal objections from computer simulations, hard data, or both.

One other theory of the origin of the Moon, called the giant impact hypothesis, is the leading theory simply because all the other theories have been eliminated. But that doesn't mean it is correct, because it has not been verified by new measurements, experiments, or observations—at least not yet.

According to the giant impact hypothesis, about 4.5 billion years ago, when the Earth was young and the solar system was still sorting itself out, with planetary bodies hurtling around every which way, a rogue planet about the size of Mars collided with the Earth. The rogue struck with such force that it was completely shattered, and it knocked out a big part of the Earth, including a great deal of matter from the mantle, the layer in the Earth that's just below the crust. The Moon then formed from the Earth debris and from some of the disrupted rogue planet as well.

If the Moon rocks reflected the original composition of the Moon, we might be able to test the giant impact theory, but they do not. The magma ocean dissolved the original rock of the lunar surface, and heavier components sank before the crust formed. We need to sample the layer below the Moon's crust, the lunar mantle. The best place to do this is in the largest and deepest impact basin on the Moon, the South Pole–Aitken Basin, which is located at the south end of the Moon and mostly on the far side that faces away from Earth. It's possible that there is mantle rock there at or near the surface. This is also a good place to look for the possible Moon ice, which might be found in the bottoms of a few craters within the basin.

The South Pole–Aitken Basin is a likely target of NASA's future manned and robotic Moon missions. China and India also have launched robotic Moon probes and China is planning a manned Moon mission as well.

We want to know more about the Moon than where it came from. We also want to know more about its history. On Earth, rock layers of various ages tell us much about geological history, and the fossils within illuminate the history of life on our planet. On the Moon, there may be ancient layers of regolith, different from the regolith that is at the surface now. If they exist, they are buried under the lava plains in impact basins, and some astronomers propose that we drill through the plains in search of fossil regolith, just as we drill or mine on Earth for fossil fuel. And just as there are Moon rocks that fell to the Earth, there should be rocks that were knocked off of the Earth long ago and landed on the Moon and remain there to this day. If we can find those "Earth meteorites" on the Moon, they would tell us about the nature of the early surface of the Earth, which has been completely destroyed by billions of years of geological changes.

CHAPTER 4

The Sun

THE SUN IN THE SEVENTEENTH CENTURY

The Sun was splendid, perfect, and changeless. That, in brief, was what educated Europeans knew about the Sun at the time of Galileo, and who could question it? The Sun's splendor was self-evident: it was by far the brightest object in the heavens. The Sun was so bright, in fact, that it hurt to look at it, so if there were any alterations in its appearance or any blemishes in its heavenly perfection, no one noticed them.

Had European astronomers read ancient Chinese court records, however, they might have been aware that the solar disk was often marred by imperfections. Chinese astrologers, who were charged by their emperor to record any event in the heaven that might affect affairs of state, had noticed dark spots on the solar disk from time to time, visible when the Sun was low in the sky and dimmed enough by the Earth's atmosphere or thin clouds so that one could stare at it directly. But even had they known of these reports, European astronomers would have shrugged them off. Given what they had learned in school, it was easy to explain the strange spots as passing clouds, puffs of smoke from terrestrial fires, or simply quirks of vision—events unconnected with the Sun itself. Conventional wisdom maintained that the Sun was a gleaming disk, as pure and constant as anything in nature. And that was that until the telescope came on the scene.

Though Galileo wrote nothing about the Sun in *Sidereus Nuncius*, his work encouraged others to build their own telescopes, and it was not long before a legion of new instruments began to reveal that the disk of the Sun was not all perfection and light. In England, Thomas Harriot saw black spots on the Sun in December 1610, and a few months later, in Germany, Jesuit astronomer Christoph Scheiner observed them, too. In a pamphlet reporting his own solar observations of 1611, Scheiner's compatriot, Johannes Fabricius, recalled how astounding it was to see blemishes on the supposedly perfect solar disk. "I directed my telescope at the Sun," he wrote. "I was unexpectedly shown a black spot whose magnitude compared with the body of the Sun was not inconsiderable . . . not believing my own eyes, I called my father," who agreed that the Sun did indeed seem spotted. Was this a hallucination? Or perhaps a temporary defect in their telescope? Neither Fabricius nor his father found it easy to sleep that night, so eager were they to see if the spot was real. Would it still be there the next day? "To my great joy," Fabricius reported, "I found the spot again, straight away. However the spot seemed to have changed its position a little, which caused some anxiety. . . ."

By the spring of 1611, Galileo himself was showing sunspots to nobles and clergy in Rome. To a modern astronomer, this seems like dangerous business, but telescopes and binoculars of today, even economy models, are much more powerful than Galileo's instruments. The concentrated sunlight they transmit through the eyepiece can cause severe eye damage. Not so with Galileo's early telescopes which, according to some scholars, were weak enough to allow him to look straight at the Sun. In addition, Galileo was careful to observe the Sun directly only at sunset, when its brilliance was considerably diminished by clouds and dust in the atmosphere. (Even at sunset, though, Galileo was flirting with the danger of severe eye damage; don't try this at home!)

Several months after he began his solar observations, however, one of Galileo's students, Benedetto Castelli, discovered a way to observe the Sun that allowed one to see it safely even when it was high in the sky. The method, called "eyepiece projection," simply involves mounting a piece of white cardboard or stiff paper a few inches or more beyond the telescope eyepiece as a screen on which the image of the Sun can be projected. Not only can the Sun be observed safely at any time of day by this method, but it is easy to sketch the sunspots themselves on the white surface, thus producing a permanent record of the positions and appearances of the spots.

Once he adopted this convenient method, Galileo began to observe the Sun more frequently and more systematically. His notebooks contain numerous sketches of the changing face of the Sun, and by comparing the sketches made on consecutive days, he clearly saw that sunspots could appear spontaneously on the otherwise unmarked solar disk and then dissipate into nothingness just as unpredictably. Moreover, while they were visible, the spots seemed to move slowly over the Sun, taking about 15 days to move from its westernmost edge to its easternmost edge.

By 1612, Galileo was convinced that the spots were real markings on the Sun and that the splendid orb of the Sun was neither perfect nor changeless. He wrote to his patron, Grand Duke Cosimo de' Medici, "Having made repeated observations I am at last convinced that the spots are objects close to the surface of the solar globe, where they are continually being produced and then dissolved, some quickly and some slowly; also that they are carried round the Sun by its rotation, which is completed in a period of about one lunar month. This is an occurrence of the first importance in itself, and still greater in its implications."

The "greater implications" of which Galileo speaks were that the teachings of Aristotle, who insisted that the heavens were perfect and unalterable, were flat wrong. This threatened the conventional wisdom of scholars, who regarded Aristotle as the ultimate authority on nature. But it also spelled trouble for the Catholic Church, which relied on Aristotle to underscore the validity of its worldview. The Church saw the world of man as corrupt and confused, and the celestial realm as changeless, perfect, and holy. Though he died three centuries before the birth of Christ, Aristotle also taught that things close to the Earth were changeable, while anything in the heavens was perfect. In effect, Aristotle provided scientific substantiation for the teachings of the faith. To challenge Aristotle, therefore, was to challenge both the natural order of things and the established authority of the Church.

It was, therefore, not surprising that Galileo met immediate resistance from more conservative astronomers. The Jesuit Christoph Scheiner published a series of letters acknowledging that sunspots did exist but arguing that they were not actually located *on* the Sun. Sunspots, according to Father Scheiner, were objects moving between the Sun and the Earth—small planets perhaps—that blocked out the light of the Sun behind them. Thus the Sun itself remained changeless.

Galileo's reply to Scheiner, published in 1613 in a series of letters called *Letters on Sunspots*, is a model of the scientific method. In it, Galileo presents the case for the spots being on the Sun with the skill of a trial lawyer. The evidence is clear, he argues: Sunspots appear to move slowest when they are at the edge of the Sun and fastest when they are in the middle of its disk. They also appear slimmer, foreshortened, when they are at the edge of the Sun, and larger when they are in the middle of the disk. What is more, when a large number of sunspots are visible at one time, they all seem to move in the same direction at the same rate across the solar disk. "To see twenty or thirty spots at a time move with one common movement is a strong reason for believing that each does not go wandering about by itself, in the manner of planets going around the Sun," he wrote. But all of these effects are to be expected if the sunspots are simply markings on the surface of a rotating sphere.

In the brief few months he had been observing sunspots, Galileo seems to have noticed almost all the important characteristics twenty-first-century textbooks attribute to sunspots, and it is remarkable for a modern reader to see how his observations went right to the heart of the phenomenon. Once formed, Galileo noted, sunspots exist for as few as 2 days or as many as 30 or 40. They are irregular in shape, and their shapes continuously change. They vary in intensity, sometimes appearing as dark as pitch, at other times appearing only as smoky smudges. Sometimes they divide into three or four spots; sometimes several spots congeal into one. They seldom appear near the poles of the Sun; most are found within about 30 degrees of the Sun's equator.

For all his insight, Galileo was not sure whether the dark dots were actually on the Sun itself or why they behaved as they did. They may, he suggested, be like clouds hovering just above the Sun. After all, don't thick clouds on the Earth appear darker than the sky surrounding them? And can they not appear and disappear as spontaneously as sunspots seem to do? Or split into several? Or merge into one? So if Galileo was not the first to report the spots on the Sun, and if his observations left him more puzzled than enlightened, his solar observations are still landmarks in the history of astronomy. Galileo was the first person to systematically describe the appearance and motion of sunspots and to draw the most important conclusions from these observations: *the Sun was changeable, and the Sun rotated.*

Beyond this, however, Galileo's observations of the Sun raised a whole host of new questions that Aristotelian physics had not even considered previously. What was the Sun made out of? How hot was it? What made it shine so brightly? Why did it rotate? What made the spots darker than the rest of the Sun, and why did they appear and disappear so capriciously? These were questions that could not simply be answered with the meager knowledge of the day, nor could they be resolved using the simple telescopes that Galileo had at his disposal. Yet it is precisely because his work raised such questions that Galileo's solar observations marked such a watershed in science. Prior observers could only marvel at the magnificence of the Sun. Galileo showed that the Sun's behavior was guided by physical laws—that its puzzles could be unlocked with the aid of ingenious instruments and the power of human reason. His observations pointed the way to centuries of fruitful research in the future. The Sun, after Galileo, was no less splendid and no less remarkable than it had been before. But it was now an object that could be understood as well as admired from afar.

THE SUN TODAY

Galileo concentrated on studying sunspots and learned that the Sun turns. Beyond that, we don't know much of what he thought about the Sun itself. Today we know that the Sun is simply a star, but one that is so close to us that it far outshines all other stars taken together, as seen from Earth. Because it is so close and bright, the Sun is easy to study with modern instruments and telescopes, and we have a pretty good dossier on it.

We also know that the Sun turns in very peculiar ways, which proves that it's not a solid body. The Sun rotates once on its axis every 25 days and 16 hours *at the equator*, just as the Earth turns once every 24 hours. But note those qualifying words *at the equator*. The Earth's rotation period is 24 hours everywhere on the planet. But the Sun turns faster near its equator than at higher latitudes toward the north and south poles. If there were a sunspot right on the solar equator and another spot at a near-polar latitude of say, 75 degrees, we would see the equatorial spot make a complete rotation in 25 days and 16 hours, while the sunspot at latitude 75 degrees would lag behind like a hopelessly outclassed competitor in the Boston Marathon, taking an additional week and then some for a complete turn around the axis of the Sun. The rotation period at latitude 75 degrees is 33 days, 9 hours, and 36 minutes.

Actually, there are few if any sunspots at the solar equator or at high latitudes; spots rarely occur that close to or that far from the equator. So there are two basic properties of the Sun that are revealed by the location and motion of sunspots: it has differential rotation, meaning its apparent or visible *surface* turns at different rates depending on the latitude north or south of the equator, and it has *sunspot zones*, or distinct regions of solar latitude where most sunspots occur.

These surface effects signal something more complicated under the surface. The Sun's rotation not only changes with latitude, but it changes with depth. The deepest part of the Sun, the core, turns faster than the surface, for example.

The Sun has no solid surface; it is completely made up of hot gas. Solar energy is generated in the core. A thick layer above the core is in constant upheaval, with hot gas bubbling and flowing up and down, like the boiling water in a kettle heated on a stove. (The upward-moving gas carries energy from below, and it cools as it rises, then sinks back down.) Somewhere near the bottom of that layer, dubbed the convection zone, there is a solar dynamo, in which the streaming of electrified gas generates magnetic fields that penetrate outward through the visible surface of the Sun, causing sunspots and extending into the Sun's atmosphere above. The surface layer, which appears to the eye as the bright disk of the Sun, is the photosphere. It takes about 170,000 years for energy from nuclear fusion in the central part of the Sun to reach the surface.

Just as you hear boiling water bubbling in the kettle, the hot gases in the Sun's convection zone produce sound waves that sweep through the interior of the Sun and cause vibrations in the photosphere that are observed by astronomers. We can see periodic changes in the brightness at different points on the photosphere, and we measure periodic changes in the velocity as areas of the surface rise and fall in response to the sound waves. We deduce the nature of the different regions within the Sun and the flow of gas streams in the interior from the measurements of these solar oscillations, just as seismologists determine the internal structure of the Earth by studying the waves that are triggered by earthquakes.

There's another and more remarkable method by which astrophysicists investigate the conditions inside the Sun. These observations are actually made deep underground on Earth. Nuclear reactions in the core of the Sun produce a steady flow of subatomic particles called neutrinos (which means "little neutral ones"). Neutrinos hardly interact with other forms of matter at all, so they can pass through huge amounts of matter with very little blockage. In fact, they can pass through the Sun from center to exterior, on through interplanetary space, and—for those that happen to strike the Earth—right through the Earth and out the other side. There are neutrinos from the Sun passing through your body at all times, whether it's day or night.

Elaborate detection equipment, typically involving many tons of water or another fluid, has been devised that can capture a tiny fraction of all the neutrinos that are passing through it. The apparatus is installed in deep mines so that the overlying rock and dirt will shield the detectors from most other atomic or subatomic particles. These other particles might trigger or interfere with the detectors, especially cosmic rays, which are high-speed atomic nuclei traveling through space, some from within the Milky Way and some from distant galaxies. These underground observatories enable us to sample the neutrinos from the region where solar energy is generated. The findings so far reveal aspects of neutrinos that were not previously known—findings that are of much interest to physicists—but they also show us that neutrinos are a telltale indicator of conditions near the center of the Sun.

We refer to the photosphere (which means "sphere of light") as the "visible surface" because there is no hard surface on the Sun. The photosphere is located just above the convection zone, which affects the structure that we see through powerful solar telescopes. Specifically, detailed pictures of the photosphere show that it is mottled by about a million seething brighter spots, each about 700 miles across—some smaller, some larger—separated by narrow, relatively dark lanes. The bright spots, or granules, are the tops of bubbles of hotter gas that have risen up through the convection zone. They cool, and the now cooler gas sinks back down at the boundaries of the granules, in the darker lanes.

We can glimpse a thin atmospheric layer just above the photosphere during a total eclipse of the Sun, when it looks like a red ring. That's the chromosphere or "color sphere." Telescopes reveal that the chromosphere is not a smooth layer but is pervaded by innumerable spurting jets of gas, called spicules, moving upward from the top of the photosphere. The region above the chromosphere is the solar corona (also seen during a total solar eclipse), and it goes on for millions of miles, thinning out into space. It's the least dense part of the Sun, consisting of the most tenuous gas, and it's by far the largest part.

The temperature of the Sun is hottest at the center, and it is cooler with each successive layer outward to the photosphere. Then, the temperature rises again through the chromosphere and into the corona. The temperature of the photosphere is just under 10,000°F, the chromosphere is around 17,540°F, depending on where you measure it, and the corona has the astronomical temperature of 1,800,000°F. We haven't figured out what heats the chromosphere and the corona and makes the temperature rise with altitude rather than drop off, but astronomers are working on it. Most likely, waves of some kind are transferring the energy upward without moving matter upward. These waves would be similar to ocean waves, which transfer energy though the water it moves through as the water stays in place, bobbing up and down as the wave passes. The waves that heat the chromosphere and the corona (if that's what is going on) are a little more complex in that they probably involve magnetic effects, and there may be different types of waves that heat each of the two layers. Another possibility is that the chromosphere is heated by waves and the corona is heated by numerous small, short-lived magnetic explosions, called nanoflares. Unfortunately for these two theories, observations of the waves don't necessarily show that they carry enough energy to heat the layers (although some experts disagree on this), and observations of the Sun are not capable of revealing events as small as the proposed nanoflares. Every year or two, NASA announces that its satellites and experts have solved the problem of what makes the corona so hot. But inevitably, questions arise, and then the experts resume working on the problem again.

SOLAR ENERGY

Like all normal stars, the Sun is powered by nuclear fusion—the merging of light atoms into heavier ones—that is underway deep inside it. It is a huge, naturally occurring nuclear reactor.

The Sun is perfectly safe if you don't get too close, you don't expose your unprotected skin to it for very long, and you don't spend a lot of time in space, where dangerous particles from solar explosions present a hazard to travelers. Nevertheless, the Sun erupts frequently in a variety of ways, and it does have considerable effects on the Earth and on the technological systems that are important to modern society.

The huge temperature and pressure near the center of the Sun make hydrogen, the lightest chemical element, fuse into helium, the next-heavier element. (It's about 27,000,000°F at the solar center.) Those nuclear reactions consume about 500 million tons of the Sun's hydrogen every second. Most of it is converted to helium, but about 5 million tons of the hydrogen per second are transformed from matter into pure energy. That's as much energy as would be released by the simultaneous explosions of 90 trillion one-kiloton atomic bombs. (The atomic bomb that fell on Hiroshima near the end of World War II had a yield of just 10 or 15 kilotons. A kiloton represents the energy equal to the explosion of 1,000 tons of TNT.) All that nuclear energy is what makes the Sun shine, and sunshine warms the surface of the Earth and makes it a good place for life.

Consider a campfire. If you don't keep adding wood to it, the fire will burn out. A coal-fired electric power plant will run down if the operators stop feeding coal into it by the trainload. The Sun is no different in concept. The mere fact that it's shining and that no one is making a monthly hydrogen delivery means that eventually the Sun will turn off. It won't even use up all its hydrogen, because the hydrogen in the outer parts of the Sun is too cool, and the pressure is too low, for nuclear fusion to occur.

Hydrogen is the most abundant substance in the Sun, and there's so much of it in the deep interior that although the Sun fuses 500 million tons of it every second, there's enough left to keep the nuclear fire going, with no perceptible change from one day to another, for a very long time. But nothing is forever. Eventually and inevitably, the Sun will peter out. You might think we would die from the cold then, but it's worse than that. First the Sun will swell up into a red giant star and shine so strongly that all oceans on Earth will boil away, and life as we know it will cease to exist. Current knowledge of how the Sun works and how its interior structure will change with time puts that event about 5 billion years in the future, so there's no need to put all your affairs in order on the Sun's account.

SUNSPOTS AND OTHER SOLAR ACTIVITY

Today astronomers recognize that the sunspots that fascinated Galileo so much are just one of many different kinds of disturbances on the Sun. These disturbances are referred to collectively as "solar activity." Galileo didn't know about these other solar disturbances, and some were not discovered until centuries after his time. Sunspots are simply the easiest form of solar activity to observe, so they are what he saw and studied with the rudimentary telescopes that he built.

Solar activity also takes the form of

* **coronal mass ejections**, in which a billion tons of solar gas erupt into space from the corona at speeds of around 1 million miles per hour. Some head toward the Earth, strike its protective environment, a kind of magnetic umbrella called the magnetosphere, and cause big problems for satellites, communications systems, and more. Some strike the magnetosphere and have little effect. Other coronal mass ejections simply miss the Earth and continue through the solar system toward interstellar space.

* **solar flares**, explosions on the Sun that can last for seconds, minutes, or hours. Within the solar corona, there are many loop-shaped structures whose shape is controlled by magnetic fields. Flares occur in these coronal loops, sometimes when they bump into each other. A "moderate-sized" solar flare in July 1996 released energy equivalent to "completely covering the Earth's continents with a yard of dynamite and detonating it all at once," according to solar physicist Craig DeForest. The flares emit bursts of ultraviolet light, X-rays, and sometimes gamma rays, all invisible to the human eye but capable of affecting the upper atmosphere of the Earth. Solar flares may interfere with radio communications, and some of them eject high-speed protons, one of the basic building blocks of atoms. The protons can damage satellites and other spacecraft or interfere with their operation, and

can give a potentially deadly dose of radiation to astronauts who are in the wrong place when they strike.

* **solar prominences**, arch-shaped structures that are seen on the edge of the solar disk, jutting up above the visible surface of the Sun. A single prominence may persist for weeks or months, sometimes erupting, sometimes collapsing, and sometimes dissolving quietly into the thinner gas of the surrounding solar atmosphere.

* **filaments**, which are dark, skinny linear features seen on the bright disk of the Sun and that sometimes erupt. They are simply prominences that are cooler than the bright photosphere beneath them.

* **solar radio bursts**, short-lived emissions of radio waves that come from clumps or streams of gas moving outward through the corona. By monitoring these radio waves, we can track the moving gas far outside the Sun toward or beyond the orbits of the inner planets.

* **active regions**, disturbed areas on the Sun where sunspots persist and flares may occur repeatedly.

That's just to name six manifestations of solar activity. These and nearly all the other kinds have one thing in common: they represent effects caused by magnetism disturbing the underlying, or "quiet," Sun. Magnetism matters on Earth if you are a scout using a compass to find your way in the woods, or a migratory bird that can sense the Earth's magnetic field and use it to find your way across the sea. (Homing pigeons find their way home by sensing the magnetic field, and when a solar-induced magnetic storm occurs on Earth, they get lost.) But most of the time, we are oblivious to Earth's magnetism.

Relating the Sun to a bar magnet, in which the magnetic poles are at opposite ends of the bar and are the places where the magnetic field is strongest, astronomers expected the Sun's overall magnetic field to be stronger at the poles than elsewhere on the Sun. However, the Ulysses space probe, which has flown nearly over the solar poles several times, discovered that the magnetic field there is no stronger than elsewhere on the Sun. Thus magnetism on the Sun is vastly different from not only a bar magnet, but from magnetized planets like the Earth and Jupiter, where the magnetic field is strongest at the poles.

The strongest magnetism on the Sun is at the centers of large sunspots, where the magnetic field can be 10,000 times stronger than the magnetic field on the equator at the surface of the Earth. (By contrast, the overall, or background, magnetic field of the Sun is only about three times the field on the Earth.) Large sunspots, which can exceed the diameter of the Earth, have a dark central part, the umbra, and a distinctly less dark surrounding portion, the penumbra. They appear dark because they are a few thousand degrees cooler than the surrounding photosphere. The strong magnetic field does not control the structure of the photosphere much beyond the outer border of the penumbra.

It's another story in the outer layers of the Sun, the chromosphere and corona. There magnetism is paramount: it dominates much of what's happening in those layers, where the solar gas is much thinner than in the photosphere. Magnetism even determines the overall shape of the corona. As the solar magnetic fields grow, change, and ebb, the shape of the corona changes from day to day and sometimes from hour to hour. Every time we look at another total eclipse of the Sun, the shape of the corona is different. When we see the pearly white corona, we are looking at light coming from the much brighter photosphere beneath it. Some of the photospheric light bounces off electrons in the corona and illuminates the region. The electrons represent a trivial portion of the matter in the corona, but they trace the heavier unseen gas, which is mostly protons. Pictures of the corona show bright lobe-shaped features and bright, long streamers extending away from the Sun. These are structures in which magnetic field lines confine the coronal gas and keep it from expanding outward, so the gas piles up, its electrons reflect light, and the features look bright. Other places in the corona are dark, and nothing is seen extending out from the Sun. These dark regions are called coronal holes. In the coronal holes, the magnetic field is not closed in loops but open and extending indefinitely far away, so the hot coronal gas can expand into interplanetary space. The gas escaping from coronal holes is solar wind, and it passes through the solar system, deflecting from the magnetic fields that protect the Earth and many of the other planets, and shaping some of the tails of comets.

The solar wind not only carries solar gas, it brings along a magnetic field from the Sun, which, anchored in the Sun as it turns on its axis, winds into a spiral pattern extending far into the solar system. The solar wind fills a huge teardrop-shaped volume around the Sun, forming a kind of bubble, the heliosphere, in the interstellar gas that pervades space between the stars in our galaxy. The boundary of this region, the heliopause, may be about three times the distance between Pluto and the Sun. The Voyager 1 and 2 space probes, which visited some of the outer planets, are on their way out of the solar system and have crossed a region called the termination shock, where the wind speed drops from supersonic speed. Inside the termination shock (closer to the Sun), the solar wind is traveling at greater than the speed of sound in the thin medium of interplanetary space, while beyond the shock, it is slower than sound. Voyager 1 crossed the termination shock in December 2004 at a distance of 94 astronomical units from the Sun. (One astronomical unit, the average distance between the Earth and the Sun, is about 93 million miles.) Voyager 2, leaving the solar system in a different direction, crossed the shock in December 2007 at a distance of 84 astronomical units from the Sun. The position of the termination shock changes with time, depending on how strongly the solar wind is blowing. Beyond the shock (but not yet crossed by our probes), the heliopause marks the boundary of the Sun's control of the gaseous medium of the solar system, but solar gravity keeps a vast number of comets in orbit at far greater distances, measured in several tens of thousands of astronomical units.

There's more to solar activity than its different forms. Solar activity also increases and weakens in striking ways. There are more sunspots in some years than in others, and that pattern repeats systematically. In other words, there is a sunspot cycle, with lots of spots at certain peak times, known as solar maxima, and very few at the bottom of the cycle, or solar minimum. Sunspots are not the only forms of solar activity that follow the cycle. Solar flares and other events also occur systematically more often at some times in the sunspot cycle and less frequently at others, roughly tracking the changing numbers of sunspots. Solar maxima are about 11 years apart, as are solar minima; thus, 11 years is the average length of the sunspot cycle. Galileo was lucky to be observing the Sun in a time of relatively high sunspot numbers. Had he been looking near a minimum of the sunspot cycle, he might never have seen a sunspot and never discovered that the Sun turns.

Sunspots don't just vary in number; they also vary in position. Near the beginning of a new sunspot cycle, when the first spots occur after a period of sunspot minimum, the spots are mostly near the high-latitude edges of the sunspot zones north and south of the solar equator. As the cycle proceeds and more sunspots occur, they tend to arise at somewhat lower latitudes, so that the band of latitudes at which spots are mostly present at any given time, or sunspot belt, moves toward the equator during a single sunspot cycle. When the time of sunspot minimum is near, the last spots of the solar cycle tend to be closer to the equator.

If you think all that is complicated, consider this: sunspots are highly magnetized and sometimes appear in groups. A typical sunspot group has a north magnetic end and a south magnetic end, which tend to be roughly at the east or west geographic (or, more accurately, heliographic, or pertaining to the surface of the Sun) extremities of the group. ("North magnetic" refers to the direction that a compass needle would point if there was a compass on the Sun.) During one sunspot cycle, the spots that lead the way across the visible surface of the Sun within one group, called the leading spots, have generally north magnetic polarity in sunspot groups that are north of the solar equator, while sunspot groups that are south of the equator are just the opposite: their leading spots will have south magnetic polarity.

Then, in the next 11-year sunspot cycle, there is an opposite pattern, with leading spots in groups north of the solar equator tending to be of the south magnetic persuasion, and vice versa for groups south of the equator. So there is a more comprehensive cycle of changes in sunspots, which takes about twenty-two years.

Solar activity affects life on and around the Earth in both negative and positive ways. It stimulates the bright displays of northern and southern lights, the aurora, fascinating viewers and driving tourism to Norway, Alaska, and other good viewing areas. It also menaces people traveling in space and can even give an undesirable radiation dose to passengers and crew on high-altitude jetliners, especially on polar routes. In extreme cases, solar activity can deliver life-threatening radiation to astronauts traveling to the Moon or beyond. It hasn't happened yet, but it's just a matter of time if NASA is going to send people back to the Moon and even on to Mars. So engineers are studying how to shield astronauts who may become exposed to the dangerous consequences of storms on the Sun.

We know the Sun is not the perfect, changeless object early seventeenth-century philosophers believed it to be. Since Galileo's time, we have made huge advances in our knowledge of the Sun's structure, energy, and activity. But there is still much to learn. We know the nuclear processes that occur in the center of the Sun, but what heats its outermost layer? We know solar flares can deliver lethal amounts of radiation, but how do we protect astronauts from them? In the twenty-first century, astronomers and physicists are investigating these other questions with—in the spirit of Galileo—an open mind and a passion for discovery.

CHAPTER 5

Jupiter

JUPITER IN THE SEVENTEENTH CENTURY

Before the advent of the telescope, about all that one could say about Jupiter is that it was a bright pinpoint of light that moved from night to night among the constellations. Seventeenth-century astronomers called anything that wandered among the constellations a "planet" (the word comes from ancient Greek, meaning "wanderer"). By that definition both the Sun and the Moon were planets. So were Mercury, Venus, Mars, Jupiter, and Saturn, making seven in all. (The seven days of the week, traditionally, were named for the seven planets.) Jupiter was one of the brightest of the planets in the nighttime sky—only the Moon and Venus were brighter.

Beyond that, all was speculation. The consensus of scholars, based on the teachings of Aristotle and Ptolemy, was that all the planets moved around the Earth in circular orbits, though in the mid-1500s, a Polish clergyman named Copernicus had advanced the unconventional notion that the Sun was the center of the planetary system. In addition, medieval astronomers assumed that all the planets, like the Sun and the Moon, were spherical in shape and composed of some sort of silvery, heavenly substance quite unlike the materials that made up things on Earth. If they bothered to consider the characteristics of planets, astronomers indiscriminately mixed physical descriptions with vague terms from astrology, for the distinction between fortune-telling and hard science was not as sharply drawn as it is today. Bartholomew the Englishman, writing in 1601, described Jupiter as "benevolent, hot and wet, diurnal and masculine, and temperate in its qualities. In color, it is silvery, bright, clear, and smooth."

Truth to tell, when Galileo looked at Jupiter with a telescope for the first time on January 7, 1610, he was not initially astounded by what he saw. There was Jupiter gleaming brightly in the middle of his field of view. The telescope had made it more brilliant, to be sure, and expanded it from a featureless point of light into a luminous disk, but it did not show any notable features or markings. Under the high magnification of Galileo's eyepiece, it may have shimmered a bit, like a globule of quicksilver, as the Earth's moving atmosphere bent the light rays traveling from the planet.

Still, there was something about Jupiter that January night that seemed strange to Galileo. "Beside the planet," he wrote in *Sidereus Nuncius*, "there were three starlets, small indeed, but bright. Though I believed them to be among the host of fixed stars, they aroused my curiosity somewhat by appearing to lie in an exact straight line. . . and by their being more splendid than others of their size. Their arrangement with respect to Jupiter and each other was the following:

East   *   * O   *   *West*

. . . I paid no attention to the distances between them and Jupiter, for at the outset I thought them to be fixed stars, as I have said. But returning to the same investigation on January eight, led by what, I do not know—I found a very different arrangement. The three starlets were now all to the west of Jupiter, closer together, and at equal intervals from one another as shown on the following sketch:

East   O   *   *   *   *West*

". . . I began to concern myself with the question how Jupiter could be east of all these stars when on the previous day it had been west of two of them."[7]

The lineup of the little stars was what first caught Galileo's eye, but the motion of Jupiter was what puzzled Galileo the most. In January 1610, Jupiter was moving *westward* among the fixed stars, tracing out what astronomers called a "retrograde" loop in the sky. Yet with respect to the bright little lineup of stars nearby, it seemed to be moving *eastward*. So Galileo looked again, fighting off impatience the next night as a cloudy sky prevented him from seeing anything at all. The night after that, on January 10, the clouds parted, Jupiter appeared, and the mystery was solved. The stars were still in a straight line, but Jupiter appeared to have turned around and moved *westward* again! Galileo was quick to come to an understanding: "my perplexity was now transformed into amazement. I was sure that the apparent changes belonged not to Jupiter, but to the observed stars, and I resolved to pursue this investigation with greater care and attention."

On the following nights, he watched with growing assurance as the little stars appeared in various arrangements and various distances from the planet. Sometimes only two were visible, as if one were blocked from view by Jupiter. Sometimes four were visible—it appeared, after awhile, that there were actually four stars, but that only three had been visible on the first few nights he observed. Sometimes the stars were close to Jupiter and to each other; other times they were spread far apart—still roughly in a straight line with Jupiter. It was clear, when all was said and done, that the four stars were orbiting around Jupiter itself. The one closest to Jupiter orbited in the shortest time, completing one trip around Jupiter in a little less than two days. The most distant of these little stars took a little over two weeks to go around.

---

[7] Galilei, Galileo. *Discoveries and Opinions of Galileo*. Stillman Drake, trans, New York: Anchor, 1957. 51–52.

Galileo had, of course, discovered the four largest moons of Jupiter, which today are called the Galilean satellites in his honor. But in his time, the notion of a moon going around anything other than the Earth was a new and absolutely remarkable idea. Everything in the universe, it was taught, revolved around the Earth, which was at the center of the universe. Aristotle's laws of physics required it to be so, and critics of Copernicus's Sun-centered theory had protested that the Earth simply couldn't go around the Sun, since heavenly bodies by nature went around the Earth. What is more, said the critics of Copernicus, if the Earth went around the Sun, it would surely leave the Moon behind—and even Copernicus agreed that the Moon was in orbit around the Earth.

Though he had not yet publicly committed himself to Copernicus's theory, Galileo was quick to see the implications of his observations of satellites around a planet. "Here we have a fine and elegant argument for quieting the doubts of those who, while accepting with tranquil mind the revolutions of the planets about the Sun in the Copernican system, are mightily disturbed to have the Moon alone revolve about the Earth and accompany it in an annual rotation about the Sun . . . now we have not just one planet rotating about another while both run through a great orbit around the Sun; our own eyes show us four stars which wander around Jupiter as does the Moon around the Earth, while all together trace out a grand revolution about the Sun."

Of all the revelations of his new telescope, Galileo believed that the discovery of Jupiter's moons was by far his most important. It surely convinced him, if he had any lingering doubts, that Copernicus's Sun-centered system was not only elegant and plausible, but also in accord with the laws of nature. Two decades later, he would write an eloquent defense of the Copernican system in his *Dialogue on the Great World Systems*, but in 1610, it was enough for him to point out that the Earth was clearly not the only center of motion in the universe, and to claim credit for the first discovery of new worlds in the history of mankind. That surely was something to be proud of.

Accordingly, Galileo announced the discovery of the little stars around Jupiter prominently on the title page of *Sidereus Nuncius*, which promised "Great, Unusual, and remarkable spectacles . . . in the Surface of the Moon, in innumerable fixed stars, in Nebulae, and above all in FOUR PLANETS swiftly revolving about Jupiter at differing distances and periods, and known to no one before the Author recently perceived them and decided that they should be named THE MEDICEAN STARS."

Naming the moons, however, turned out to be a more difficult proposition than discovering them. Galileo wanted to immortalize the remarkable four by naming them after his patrons, the Medici, both as a sign of gratitude for the financial support he got from the family and as a way of ensuring that the support would continue (and, hopefully, increase). But until the twentieth century, when the International Astronomical Union came into existence, there was no recognized organization that sanctioned the naming of new objects in the heavens. Galileo's discovery, in fact, represented the first time in recorded history that anyone had discovered anything new and permanent in the heavens that needed to be named! In effect, one could call the new moons anything one fancied—and astronomers were slow to agree on what names to attach to them.

Those of other nationalities, not beholden to the Medici family, looked for alternatives to the "Medicean Stars." Dutch astronomer Simon Marius, who observed the moons independently around the same time as Galileo, preferred to name them after *his* patron, the Duke of Brandenburg. Marius claimed to be the real discoverer of Jupiter's satellites, and while he may have actually seen them a few days before Galileo, scholars generally agree that Galileo was the first to figure out that they were moons orbiting Jupiter. In any case, neither "Medicean Stars" nor "Brandenburg Stars" gained universal acceptance.

In the 1600s, and over 200 years following, most astronomers simply referred to the satellites by number, starting with the closest satellite, Jupiter I, and moving outward to Jupiter IV. The names we use today were not adopted universally until the late 1800s, when additional moons began to be discovered, some of them closer to Jupiter than Galileo's original four. Ironically, it was Simon Marius who proposed their current names as an alternative to the "Brandenburg Stars." It was the German astronomer Johannes Kepler who suggested to him—perhaps half in jest, according to Marius—that the four moons should be named after mortals who, according to Greek mythology, had been pursued and seduced by the god Jupiter: three young maidens, Io, Europa, and Callisto, and one young boy, Ganymede.

Later, the notion of mythological names appealed to the classically trained Victorians, even if the licentious behavior of Jupiter might have caused some hesitation. And so, the mythological names for the moons began to gain acceptance in textbooks and technical reports in the late 1800s. Io, Europa, Ganymede, and Callisto are now the names used universally for the four Galilean satellites, starting with the closest to Jupiter, and the many other moons of the planet that have been discovered in recent years are, by convention, named for other mythical lovers of Jupiter. Fortunately for astronomers there were plenty of those.

Jupiter itself kept its mysteries from Galileo, despite the power of his clever new telescope to make it brighter. It was unresolvable through the eyepiece, a blob of light as featureless as the four new moons that circled it. As telescopes improved, details of the planets gradually began to emerge, like faces emerging from a fog. Other astronomers in the later half of the 1600s were able to see the planet as a distinct disk and could make out colorful markings, dark bands, and mysterious spots on its surface (though the moons themselves remained essentially unresolvable). Better telescopes, however, couldn't see much more. Though Jupiter turned out to be the largest of the planets circling the Sun, there was a limit to the level of detail that could be seen from the distance of the Earth, and eventually, despite improved instrumentation, it seemed as if astronomers were reaching a point of diminishing returns. Until the dawning of the space age, Jupiter and its retinue of satellites remained almost as enigmatic as they were when Galileo first gazed upon them.

JUPITER AND ITS MOONS TODAY

Today we know Jupiter as the largest planet in our solar system. With 318 times as much mass as Earth, it is more massive than all the other planets of the solar system put together, whether you count Pluto as a planet or not. Jupiter is a source of heat and strong magnetism, too, 11 times larger than Earth and turning much faster.

Galileo discovered that Jupiter was the center of a four-moon system. Now we count more than five dozen moons orbiting Jupiter. Galileo's four moons are by far the largest. Jupiter has spots, like the Sun, but they are not magnetic disturbances like sunspots. They are weather systems that resemble huge hurricanes. And it has a ring, like the rings of Saturn but much less spectacular.

The resemblance of Jupiter and its moons—the Jovian system—to the solar system of planets centered on the Sun is hardly coincidental. Some of the same forces and processes that caused planets to form around the Sun may have given birth to at least the larger moons around Jupiter. And like the Sun, Jupiter is an energy source, although much less intense. Although it shines brightly in the reflected light of the Sun, Jupiter also emits infrared radiation, and it sends 1½ times more energy into space than it receives from the Sun.

Jupiter's internal heat may be generated by the gradual contraction of its interior caused by gravity, a process that began when the planet formed 4.5 billion years ago and may still be under way. In any case, Jupiter doesn't generate energy by nuclear fusion, the way the Sun and other stars do, because it's not massive enough to produce sufficiently high temperatures and pressures near its center. Thus, astronomers sometimes refer to Jupiter as "a failed star."

Jupiter's powerful magnetic field is over 10 times stronger than Earth's and extends much farther into space. The Earth's magnetism is generated by currents flowing in molten iron deep inside our planet. Jupiter has another source of magnetism: at the high pressure at great depths inside the giant planet, hydrogen (the most common chemical element) is transformed from a gas to a metallic fluid that conducts electricity just as iron does. The region of metallic hydrogen works as a natural dynamo as Jupiter spins on its axis once very 9 hours and 56 minutes to generate the Jovian magnetic field.

Jupiter's magnetic field traps great numbers of electrons, protons, and other electrically charged atomic and subatomic particles in a large region around the planet, the Jovian magnetosphere. Jupiter's magnetosphere is larger than that of Earth or any other planet in the solar system. It is about 100 times larger than the planet itself. To appreciate that dimension, consider this: When you view the full Moon, it looks much larger than Jupiter, because Jupiter is much farther away. But if you could see Jupiter's magnetosphere (it's invisible, so you can't, but it has been mapped with radio telescopes and with instruments on space probes), it would look much larger than the Moon. And the Jovian magnetosphere sports a long magnetic tail pointing away from the Sun. It reaches more than the 405 million miles from Jupiter to the orbit of Saturn, the next planet from the Sun.

Spacecraft that venture into the Jovian system are exposed to intense radiation by the particles in Jupiter's magnetosphere. It's enough to damage electronic components, and in places, the radiation dose would kill anyone who ventured near.

There's no surface on Jupiter; the gas giant has a huge atmosphere with bands, zones, and jet streams. There are powerful lightning flashes and, near the north and south poles, aurora like the Northern and Southern Lights on Earth. What we see through a telescope are the clouds—some pale, some colorful—high in the atmosphere. Many are organized in large rotating systems, like hurricanes or cyclones. Some storms and cloud features come and go, but others persist. The largest and most famous, Jupiter's Great Red Spot, is oval, almost as wide as the Earth, and with its long dimension almost twice the size of the Earth. It is a hurricane that has lasted for at least 130 years and perhaps much longer.

The four moons that Galileo discovered are each a different world, and to many astronomers, ourselves included, they are the big story in the Jupiter system. The smaller two, Io and Europa, are each about the size of our Moon. The larger two, Ganymede and Callisto, are each about as big as the planet Mercury. In fact, Ganymede is the largest of the 168 moons in the solar system that were known as of early 2008, and the Jovian system, with 62 known moons, accounts for more than one-third of that total.

In our solar system, the planets near the Sun—Mercury, Venus, Earth, and Mars—are solid bodies of rock and iron. In contrast, the giant planets—Jupiter, Saturn, Uranus, and Neptune—have huge atmospheres that extend to great depths, farther than we can see or reach with space probes. And they are much lighter (less dense) than rock.

The four Galilean moons follow a similar pattern, with hotter, smaller, denser, and rockier Io and Europa on the inside—closer to Jupiter—and colder, lighter, larger, and icier Callisto and Ganymede farther from the planet.

FOUR GREAT MOONS

Io is the smallest of the moons that Galileo found but the most remarkable and dynamic one. It teems with volcanism to a degree unlike any other known body. Volcanoes mark its surface, some filled with magma seething at more than 3,000°F, hotter than in any volcano on the Earth today. Together, they produce 100 times as much lava as all the volcanoes on Earth. Heat radiating into space from Io, per square inch, is 30 times greater than the heat flow from Earth. The surface of Io is marked by splotches of sulfur-bearing compounds from the volcanic emissions. If you imagined Hell as medieval artists portrayed it, Io would not be far off the mark. Io's present dreadful or awesome conditions (depending on how you feel about them) resemble the high-temperature volcanic activity that geologists believe prevailed on the early Earth, until about 3 billion years ago.

There are over 200 volcanic craters that are more than 12 miles wide on Io today, compared to only about 15 on all the Earth, even though Earth is much bigger. The largest Io volcano, Loki Patera, is more powerful than any on Earth. It's wider than the whole Big Island of Hawaii, with its famous Mauna Loa, Mauna Kea, and Kilauea volcanoes put together. By one estimate, Loki puts out 10 trillion watts on a typical active day, like the mother of all hair dryers.

Plumes of volcanic emanations, consisting of gas and tiny solid particles ("dust"), sometimes reach nearly 300 miles above the surface of Io. Some of this gas escapes and forms a donut-shaped cloud around Jupiter, called the Io torus. Dust ejected from Io becomes electrically charged and driven to speeds of over 400,000 miles per hour. The high-speed dust streams out among the planets in the solar system, even passing near the Earth. Some of it must escape into interstellar space, where, if it survives long enough, some day it may enter another planetary system around a star beyond the Sun. Some of the gas erupting in the plumes becomes ionized, meaning that gas atoms lose electrons. Great numbers of these electrons flow toward Jupiter along the planet's magnetic field lines, creating a huge electrical circuit in space.

Lakes of lava at some of Io's volcanoes glow in the dark, producing hot spots on the night side of Io (or when Io is in the shadow of Jupiter) that can be seen against the moon's unlit surface in images made with cameras that work in infrared light.

Besides the hundreds of volcanoes, there are at least 100 other mountains on Io that did *not* form by volcanism. They were created by fracturing and faulting of the moon's crust, as occurs at some places on Earth. Some of them are as much as 10 miles high, while the altitude of Mt. Everest is less than 6 miles.

Beneath its rocky, sulfurous crust, Io has a mantle of molten rock and a heavy iron core. It spins on its axis every 42.5 hours and speeds along in its circular orbit around Jupiter, making one orbit in that same amount of time.

Europa orbits a bit further from Jupiter than Io. It makes one revolution around Jupiter every 3 days and 13 hours, and it, too, has internal warmth. It's not rocked by volcanic eruptions by any means—there are no volcanoes—but it has a moon-wide subsurface ocean of salty water, sandwiched between a surface layer of ice and an underlying rocky mantle. A few warm places, or "hot spots," dot the ice crust. The water may react with minerals in the rock below, and in places it may be heated partly by internal volcanic activity, like the hot vents on the sea floor of Earth. Some biologists believe that these conditions are suitable for primitive life to arise by natural means. Below the rock, Europa has an iron core.

In some places, the icy crust of Europa breaks up into large ice rafts, typically a few miles long, reminiscent of ice floes in the Arctic

Ocean on Earth. There are also long ridges in the ice, pits, and small rounded formations called domes. All these features indicate that the ice layer is disturbed and warped by internal forces a bit like those at work in the crust on Earth and probably by upwelling of relatively warm water from below. Reddish and brown hues in the surface ice may be caused by sulfur ejected from volcanic plumes on Europa's neighbor, Io, and by minerals brought up by the rocky mantle overturning or currents in the ocean.

The total thickness of surface ice and subsurface ocean on Europa is probably about 90 miles, but experts disagree about how far below the surface the ocean begins. If the ice were thin, some speculate, it might be possible to insert a small robotic submarine, brought from Earth onboard a spacecraft, to sample the hidden waters and search for traces of life.

Ganymede is very icy indeed. Its frozen crust and the icy mantle just below it together extend more than a quarter of the way to the center of the moon. The surface of Ganymede is marked with signs of stretching and fracturing. Some parts are dark and heavily cratered from the impacts of smaller bodies, meaning that they have persisted at the surface for aeons. (The more impact craters there are on an area of a given size, the longer it has been bombarded. Surfaces that are relatively crater-free have been resurfaced by one or another geological process, erasing the old craters.) In some places, however, the dark terrain is crossed by bright material that may be relatively fresh ice that has come up from below the surface.

There are warmer regions inside Ganymede where all is not frozen: There is a salty ocean with ice above it and ice below. Beneath the ice and water is a rocky mantle and perhaps an iron core. Sealed in ice with perhaps no contact with minerals in the rock that might provide nutrients or chemical building blocks, the Ganymede ocean seems a less likely setting for life than Europa's buried sea. It is the only moon in the solar system, as far as we know, that generates its own magnetic field, as the Earth does. The magnetism may come from a dynamo effect as currents of molten iron circulate in the core, if indeed there is an iron core and if it is at least partly molten. We have a lot to learn about Ganymede.

Callisto resembles a fractured dark and icy billiard ball. It, too, has a subsurface ocean, bordered by layers of ice above and below. It's not very dense and may have little rock or iron. There's no evidence of a dense metal or rock and metal core. The many large impact craters on its surface, which range in size up to that of the Valhalla structure, 610 miles wide, tell us that the surface is very old, dating back about 4 billion years to an early epoch of the solar system. However, for every large crater, there should be many more small craters, and there should be more of the smallest craters than any other size. That's because there are more little rocks coursing through interplanetary space than big rocks, so small craters form much more frequently. However, small craters are not as common on Callisto as expected. This means that there is a geological process at work, or which was at work in the past, that rubs out or covers over small craters, without noticeably altering the big ones.

In summary, there are immense differences and some striking resemblances among Galileo's four moons of Jupiter. Going from the inside out (from nearer to Jupiter to farther from Jupiter), they range from warm and enormously geologically active (Io and its volcanoes) to cold and largely stagnant (Callisto and its only weakly resurfaced crust). Most of the heat that drives the geological processes on these moons comes from the tidal force of Jupiter, which works on Io and Europa now and which probably affected at least Ganymede in the distant past.

Jupiter's many smaller moons range down to just a mile or so in size. As telescopes and electronic cameras improve, astronomers are finding more and more of them. Many orbit around Jupiter the same way as the Galilean moons, in the same direction that Jupiter turns. Others travel in the opposite, or retrograde, direction. They may have been captured in orbit around Jupiter, rather than forming around it like the other moons.

CHAPTER 6

Saturn

SATURN IN THE SEVENTEENTH CENTURY

Saturn was a source of abiding frustration to Galileo, a puzzle he was never able to resolve. Flush with his discovery of the four starlike moons circling Jupiter in January 1610, he hurried to publicize his observations in *Sidereus Nuncius*. But the book, which was released in March 1610, made no mention of Saturn, for Galileo had not yet observed it closely. Even though Galileo must have been curious to find out whether Saturn resembled Jupiter, he had no way of doing so. The planet was on the other side of the Sun during the early months of 1610 and could not be seen easily in the nighttime sky for several months thereafter. Like it or not, Galileo had to wait for the turning of the seasons and the realignment of the heavens before he had a chance to inspect Saturn with the new instrument. If he was impatient to beat other astronomers to the punch on Saturn, he could at least be confident that, before July of 1610, no one on Earth could see the distant planet any better than he could.

On July 16, he was finally able to spy it, high in the evening sky. Gazing through the eyepiece, he was so astounded by Saturn's magnified image that he could scarcely contain himself. After several nights of further observations, he dispatched an excited letter to his patron, Cosimo de' Medici. "I discovered another very strange wonder," he wrote. "This is that the star of Saturn is not a single star, but is a composite of three, which almost touch each other, never change or move relative to each other, and are arranged in a row along the zodiac, the middle one being three times larger than the two lateral ones, and they are situated in this form oOo. . . ."[8]

Galileo was eager to share the discovery with other scientists, too, but he was afraid that if he indiscriminately announced that Saturn was a triad, others might observe it, too, and claim that they had seen it first—apparently not an uncommon occurrence in those times. And so he sent letters to his scientific correspondents with the new discovery hidden in an anagram, the letters scrambled into a meaningless string of syllables: SMAISMRMILMEPOETALEUMI-BUNENUGTTAUIRAS. No one was likely to decipher it, reasoned Galileo, and if anyone were to claim priority in the matter, Galileo could present the anagram and its solution, dated and signed by himself, as proof that he had made the discovery first.

A few months later, after he was sure that he had established his claim to having observed Saturn's secrets through a telescope, he sent a follow-up letter with the solution to the anagram: *Altissimum planetam tergeminum observavi*," which is Latin for "I have observed the highest planet [Saturn], triple-bodied." The great German astronomer, Johannes Kepler, who received the letter, actually managed to make some sort of astronomical sense out of the anagram, decoding it incorrectly as "*Salve umbistineum geminatum Martia proles*," which means "Hail, burning twin, offspring of Mars." Kepler, a clever mathematician with a love of puzzles, thought that Galileo had found two moons around Mars!

Kepler's decoding of the anagram made sense to him, because he had already predicted that Mars would have two moons—since the Earth had one, and Jupiter four, it seemed mathematically obvious to

---

[8] Van Helden, Albert. "Saturn and His Anses." *Journal for the History of Astronomy*. Volume 5, 1974. 105.

him that Mars would have two, and Saturn eight, forming a nice geo-metrical progression. And so Kepler interpreted Galileo's message to confirm what he already suspected, though it soon turned out that he was way off the mark. Mars does, indeed, have two moons, but they are so tiny that Galileo could not have seen them at all. They were not discovered until the late 1800s.

Galileo was also guilty of seeing only what he expected to see. Excited by his discovery of four moons around Jupiter, he inter-preted the indistinct image he saw through his telescope as a large central planet with two moons around it, when, in fact, what he saw was Saturn and the two "handles" of Saturn's rings, sticking out on either side of the planet.

Although he observed Saturn on a number of occasions during the rest of his life, Galileo was never able to interpret the companions of the planet as anything other than moons. The very notion that a planet could be girdled by a vast flat ring never seemed to cross his mind, even when he was staring right at it.He could see with his own eyes that Jupiter and Venus were spheres, as were the Moon and the Sun, and scholars at the time generally agreed that *everything* in the heaven was spherical. There was no reason to believe that Saturn was any different.

Still, Saturn's odd appearance didn't seem to make sense, and the more one looked through the telescope, the more mysterious it became. Other astronomers reported to Galileo that Saturn appeared to them not as three objects, but as a single oval or even a rectangle. Galileo advised them that the fault was in their telescopes: sufficient magnification and better optics would separate the two moons from the main body of the planet—and he was right in that regard. Still it seemed strange that the two moons that flanked the Saturn didn't seem to move around it like the moons of Jupiter moved around their mother planet. They just hung there in space, it seemed, doing nothing—or so it appeared at first.

Galileo continued to observe, hoping for some clue as to what was going on, and near the end of 1612, after a short break in his observation routine, he turned his telescope to Saturn once more and saw an even stranger sight than he could have imagined: the moons, unchanged in appearance for so long, had now disappeared entirely! "I found [Saturn] to be solitary, without its customary supporting stars, and as perfectly round and sharply bounded as Jupiter. Now what can be said of this strange metamorphosis? That the two lesser stars have been consumed, in the manner of the sunspots? Has Saturn devoured his children?"[9] Galileo was joking, of course, referring to the ancient myth that the god Saturn, learning of a prophecy that he would be overthrown by one of his sons, cannibalized his offspring one by one as soon as they were born. (In the 1800s, Francisco Goya, the great Spanish artist, painted one of Saturn's grisly meals in a horrific painting that now hangs in the Prado in Madrid. It is an unforgettable image.)

But Galileo was certain that the moons of Saturn had not disappeared permanently, and he was even more certain that if he could not see them, the fault was *not* in his telescope, as some skeptical critics suggested. Within a couple of years, he wrote, he felt confident that the moons would be seen again. His reasons involved a complex geometric arrangement in which the orbits of the moons sometimes carried them into a position where they were no longer illuminated strongly by the Sun, only to move into the light again after a while. The idea was ingenious, but Galileo was not able to spell out the details very clearly, and as we know today, his explanation was pure fantasy. Yet as if to prove him right, the moons *did* reappear a year or so afterward.

What was making the rings disappear, in fact, was that about every 15 years (as Saturn orbits the Sun once in nearly 30 years), its rings, which encircle the planet directly around its equator, come into alignment with the line of sight to the Earth and are seen edge-on. At these times, because the rings are very thin, they seem to disappear, just as the thin blade of a razor seems to disappear when one looks directly at the edge of the blade. For a few months the rings are not visible, and then, as Saturn continues on its orbit, they begin

---

[9] Galilei, Galileo. "Letters on Sunspots." In Drake's *Discoveries and Opinions of Galileo*. 143–144.

to open up to view again. The relatively indistinct images Galileo saw though his telescope, even at the highest magnification, made it hard for him to recognize this—most of the time he either saw the supposed moons or he didn't.

Under favorable viewing conditions in 1616, however, Galileo noticed that Saturn had morphed again. This time, as he noted in a letter, the two moons "are no longer two small, perfectly round globes, as they were before, but are at present much larger bodies, and no longer round, as seen in the adjoined figure ⬮⬯⬮. It seems that Galileo almost had it right now, but he still did not perceive that he was seeing a ring. Instead, he described what he saw as two elliptical arcs on either side of the planet with dark triangles in the middle of each. The middle globe, he remarked, always appeared perfectly round, while the flanking moons seemed simply—and mysteriously—to have changed their forms.

For the rest of his life, Galileo continued to believe that Saturn was a triple planet consisting of a round central planet flanked by two shape-shifting brothers that appeared in three forms: little stars that hung motionless along a line for years at a time, then disappeared, then reappeared as "ears" sticking out from the planet. It may seem odd to the modern reader, accustomed to thinking of Saturn as a ringed planet, that Galileo never recognized what was plainly before his eyes. Yet, as Louis Pasteur wrote over two centuries later, referring to his own discoveries in microbiology, "In matters of observation, fortune favors the prepared mind." Galileo's mind was prepared, all right, but unfortunately it was not prepared for what he was seeing. At first, prepared perhaps by his reading of Copernicus, he had been willing to abandon an Earth-centered view of the universe when he saw four moons orbiting Jupiter. Afterward, facing evidence that not everything in the Universe was spherical, he could only see things in terms of what he had already discovered. If Jupiter's companions were moons, then so were Saturn's.

After his first few years of telescopic activity, Galileo's zest for nighttime activity waned, and he became more involved in controversies over his unconventional views on astronomy and physics. He viewed the heavens only infrequently, and in his later years, confined to his home by the edict of the Church, he became blind, relying on letters from other astronomers for news of what was going on with Saturn and the other planets. Though he could never figure out exactly why Saturn behaved so strangely, he felt sure that, as astronomers observed it more frequently and with improved telescopes, it would be come clear at what points in its orbit Saturn changed its shape, and the mystery would eventually be solved. There would, he wrote to a friend, be "people who have the curiosity to do what I have done for so long, if not better."

He was right. After Galileo's death, astronomers continued to observe and sketch the changing appearance of Saturn, using bigger telescopes and improved optics. Before the seventeenth century was over, the Dutch astronomer Christiaan Huygens had made the conceptual leap that eluded Galileo, recognizing that the shape-shifting of Saturn was easily explained if its equator was encircled by a gigantic flat ring. With this understanding, Galileo's two companion moons were no longer necessary—Saturn was now perceived as we know it today, as the ringed planet.

Yet Galileo was not completely wrong about Saturn's moons— the planet does not travel alone through space. In 1655, Huygens also discovered that Saturn had a moon that orbited it just like the four moons of Jupiter. He called it Titan, and it was just the first of many moons which were eventually discovered around the planet. When Saturn was visited by space missions centuries later, it turned out that the real moons of Saturn were far more remarkable than the fictional pair Galileo persisted in seeing for so many years.

SATURN TODAY AND PLANETS BEYOND

Today we know Saturn as the second-largest planet in the solar system, just a bit smaller in size but much less massive than Jupiter. It was the most distant planet known in Galileo's day. But astronomers have now explored Uranus and Neptune, two large planets far past Saturn, as well as little Pluto (currently called a dwarf planet or an "ice dwarf") and others like it. And we have found many large planets, so-called exoplanets, that are dozens of light-years or more from Earth, orbiting other stars beyond the Sun. (One light-year, a unit of astronomical distance, is almost 6 trillion miles.)

Saturn is the lightest (least dense) planet in the solar system; much of it is hydrogen, the lightest chemical element. It's adorned with clouds of water ice particles, ammonia ice, and other substances. There's a fast wind at its equator, racing east at 1,100 miles per hour. Roughly every 30 years, which is about once per Saturn year, prominent white clouds form at a spot in the northern hemisphere when summer is underway. The clouds expand as the wind spreads them in an east–west band around the whole planet.

Deep inside Saturn, droplets of liquid helium continually descend toward the center of the planet, producing heat by friction. That heat may be the source of Saturn's infrared glow. Like Jupiter, it emits more energy than it receives from the Sun. You see Saturn and Jupiter by the reflected light of the Sun—that's visible light. The excess energy from these giant planets comes off of them in the form of infrared rays.

Saturn can't challenge Jupiter for the heavyweight title among planets in the solar system; Jupiter has 318 times the mass of Earth, while Saturn has just 95. Though Saturn is nearly as big as Jupiter, it's filled with lighter stuff. A chunk of matter with the same density as Saturn would float in your bathtub.

Saturn spins quickly, like Jupiter, about once every 10 hours and 47 minutes. (Jupiter's spin period, or "day," is 9 hours and 56 minutes.) Their rapid spins make the two giant planets bulge at their equators and flatten at their poles. So the diameter across Saturn or Jupiter is larger at the equator and shorter through the poles.

Saturn has lightning storms, a powerful magnetic field and a magnetosphere, like Jupiter, and dozens of moons besides large Titan. And, of course, it has the beautiful rings. When you think of Saturn, you think "rings." A key point, as we will see, is that all but the very smallest moons orbit farther from Saturn than the rings. This distinction tells us something important about the forces at work in orbit around a planet.

THE MIGHTY MOON TITAN

Titan is titanic, by far the largest moon of Saturn and ranked second only to Jupiter's Ganymede among the almost 170 known moons. But what most distinguishes Titan is its atmosphere. On all other moons, the atmosphere, if any, is thinner than a laboratory vacuum. But Titan's atmosphere is thick, cloudy, and hazy, with a barometric pressure at the surface that is half as high as the air pressure at sea level on Earth. And like the Earth, Titan has a greenhouse effect. Its thick atmosphere traps heat and makes the surface, although frigid by human standards, much warmer than it would otherwise be, given Titan's great distance from the Sun (almost 890 million miles). The surface temperature gets up to -290°F (-179°C) at most.

It's not just Titan's surface temperature that impresses us—it's what we find on the surface. Titan has hundreds of lakes. The lakes come in all shapes and sizes, and at least one, with a surface area of more than 38,000 square miles, is larger than the biggest of the Great Lakes, Lake Superior. But Superior is superior in the amount of water, because the Titan lakes have none. They are filled with liquid hydrocarbons, methane and ethane. On Earth, methane and ethane are flammable gases, but on Titan, at the high latitudes where the lakes occur, these hydrocarbons are liquid, like crude oil beneath the desert in Saudi Arabia. However, if Titan were a country, it would outrank Saudi Arabia and indeed all of the oil-producing countries in OPEC put together. The Titan lakes were mapped with a radar aboard the Cassini spacecraft. From these maps we learned that "Titan has hundreds of times more liquid hydrocarbons than all the known oil and natural reserves on Earth," according to a statement from NASA's Jet Propulsion Laboratory in February 2008.

Where the surface is a bit warmer, closer to the equator on Titan, methane evaporates and floats up as a gas. High in the atmosphere, the methane cools and precipitates, falling in drops like rain on Earth.

Scientists speak of the hydrologic cycle (water cycle) on Earth, in which rain falls on the land and runs down into lakes or the sea and evaporates, rising as water vapor up to the clouds and down again as rain. There's a hydrologic cycle on Titan, but it's a methane cycle. Geographic formations that look like dry river channels may show where the liquid flows (or did flow) when it crosses terrain outside the lakes. On Earth, fast-moving streams tumble rocks, making rounded pebbles. On Titan, there are round pebbles around the place where the European Space Agency's Huygens space probe landed. Scientists think that the chemical conditions and the abundant liquids on Titan may resemble the circumstances in the early history of Earth, when life first arose.

Titan is the only known body other than the Earth with standing pools of liquid on the surface that are not volcanic magma. But Titan is not all lake country by any means. It has deserts, too. The surface is marked by fields of parallel dunes as large as 900 by 100 miles or more that remind geologists of the Namib Sand Sea in Africa, where long ridges of sand dominate the landscape. It's not clear what Titan's "sand" is made of, but scientists have suggested that it is particles made of frozen hydrocarbons. The Jet Propulsion Laboratory, describing work of the planetary scientist Ralph Lorenz and his coworkers, says that "[t]he dark dunes that run along the equator [of Titan] contain a volume of organics several hundred times larger than Earth's coal reserves." If you voyage to Titan, you won't have to bring along a bag of charcoal or a cylinder of propane to cook your meals.

INTRIGUING IAPETUS

It's all bright versus dark on Iapetus, another of Saturn's largest moons. At least, there is a huge dichotomy on Iapetus: one hemisphere of the moon's surface is about 10 times darker than the opposite side. That's the most extreme variation between one side of a moon or planet and the opposite side that is known to exist anywhere.

It's useful to think of Iapetus in terms of two hemispheres, just as on Earth we sometimes contrast the Western Hemisphere ("New World") with the Eastern Hemisphere ("Old World"). On Iapetus, the hemispheres are distinguished by the directions that they face, not their east–west or north–south orientations. They are the leading and following hemispheres. The leading hemisphere always faces forward in the direction that Iapetus moves in its orbit. The following hemisphere always faces back in the direction Iapetus has come from. This happens because Iapetus is in synchronous rotation, meaning that it turns once on its axis (an Iapetus "day") in exactly the time that it takes to go once around Saturn. That keeps each hemisphere facing the same way in relation to the moon's orbit at all times.

NASA's Cassini probe snapped the best pictures of Iapetus that we have. They show that the dark region that takes up much of the leading hemisphere is cut in two by a high mountain ridge that runs for more than 1,200 miles along the equator. Some of these mountains are as dark as the plains 12 miles beneath them, and some are brighter, as though a thin layer of dark surface material has slid off.

It's possible that early in its history, Iapetus was spinning faster than it is now, and that it was warmer and softer. So it may have produced an equatorial bulge, just as the fast-spinning Saturn and Jupiter have done. Then, as the moon cooled, the bulge hardened in place, becoming a mountain ridge even as Iapetus turned slower and slower until it reached its present rate of once per 79 Earth days.

But what causes the big difference in brightness between the dark leading hemisphere and the bright trailing side? It may be an external effect, meaning that the dark matter is swept up by Iapetus or bombards the moon from space. Or it might be internally generated, meaning that the color difference is caused by material coming from within Iapetus. The jury is out on this one, but we favor an external source for the dark stuff. It could be particulate matter ("dust") coming from another moon of Saturn, even a small disrupted moon that no longer exists. Or perhaps it is dust knocked off of one of the other moons by falling meteoroids or some other cause. It might even be dark chemical compounds blown off the top of Titan's thick atmosphere. The dark area shows no evidence of a surface covered by lava or some other liquid that has gushed out from below and then frozen, so the internal theory of Iapetus's hemispheric differences seems less likely than an external cause.

ICY ENCELADUS

Another Saturn moon, Enceladus, is one of the most interesting and active satellites in the solar system, even though it's just 314 miles wide. It has about the same density as ice and must be largely composed of ice, but it contains liquid water as well. The water is not buried beneath miles of crust, as on Jupiter's moons Europa and Ganymede. It is probably less than about 50 feet below the surface in the area surrounding the south pole.

Most of the surface of Enceladus has a smooth, white covering, probably composed of tiny ice particles. Thanks to the icy coating, Enceladus is the most reflective known object in the solar system. It's marked by an occasional impact crater as well, and some craters are deformed as though the crust sank beneath them and then ice welled up from below. But the south polar region has few impact craters and shows much evidence of recent geologic activity, which wiped out craters and thrust up great ice boulders, some as small as a house and some as large as a football field. Four long, narrow, dark cracks in the crust there, nicknamed tiger stripes, run through "hot spot" areas where heat is flowing upward from below. And erupting in at least eight spots along the tiger stripes are jets of water vapor and ice particles. The jets merge in a plume that extends thousands of miles above the surface of the moon. Most of the ice falls back to the surface, coating the moon and making it bright. The Cassini spacecraft, which discovered these phenomena, also detected what are described as "fingerlike projections" from the plume that run all the way to one of Saturn's rings, the E ring (which we describe in the next section). The combination of a heat source and liquid water near the surface of Enceladus mark it as a possible habitat for life.

We could go on and on about the moons of Saturn, because we have described only a few of our favorites. But we won't. Saturn has 60 moons that we knew of in early 2008, and more are sure to be found. Instead we will describe the rings. But that will bring up some of the strangest moons of all—the "flying saucer moons"!

THE RINGS OF SATURN

The rings of Saturn are perhaps the most spectacular sight in the solar system. They circle the planet in an immense, flattened array, extending from about 41,600 to 300,000 miles from the center of the planet. The radius of Saturn is 37,449 miles, so the innermost ring (D ring) is just a few thousand miles above the planet.

Some of the rings break up into innumerable concentric ringlets. One ring (F) is braided and knotted and contains spiral patterns. Another, the brightest and most massive (B), is sometimes graced by so-called spokes, which seem to be clouds of very small particles that may be levitated by electrical force and which, like the spokes in a bicycle wheel, are oriented radially toward and away from Saturn. Unlike a bicycle wheel's spokes, Saturn's ring spokes come and go. And they seem to extend up and down, above and below the main body of the B ring, as well.

The rings of Saturn extend for many thousands of miles within the equatorial plane. In the direction perpendicular to the equatorial plane, the rings are, for the most part, very thin. The ones we can readily see are mostly less than six-tenths of a mile thick.

Five of the seven rings, including the brightest and most massive ones, are located between 41,000 and 87,600 miles from the center of the planet. All of Saturn's large moons, and some of the small ones, are located beyond about 93,000 miles. That's no accident. The distance of 93,000 miles is the approximate position of the Roche limit, which marks the outer boundary of a danger zone for moons.

Within the Roche limit, named for French mathematician Edouard Roche, tidal force can pull an object such as a moon apart. Tidal force is the difference between the pull of a planet's gravity—in this case, the gravity of Saturn—on the part of the moon that is closest to the planet and the part that is farthest. The bigger the moon, the greater the tidal force on it. But if a moon is sufficiently small, or made of rock that is sufficiently strong, it can survive, because the tidal force on its opposite sides won't overcome its internal strength. So small moons can exist inside the Roche limit, among the rings, and some do.

The rings that are inside Saturn's Roche limit (A, B, C, D, and F rings) consist largely of debris from ancient moons that suffered catastrophic disruption when they came too close to Saturn. The G and E rings are outside the Roche limit and are made from particles knocked off of adjacent moons by the impact of meteoroids, or even (in the case of E) from particles that erupt in geysers on the moon Enceladus. A previously unknown ring, found by the Cassini spacecraft in 2006, lies outside the F ring and inside G and E in the vicinity of the orbits of two small moons, Janus and Epimetheus. It's probably composed of particles knocked off of their surfaces.

The B ring is largely composed of pieces of ice, ranging from very tiny to as big as a small house. The two outermost rings, G and E, consist of very small particles. E is much thicker than the other rings, extending about 19,000 miles perpendicular to the equatorial plane, but it is so sparsely filled with particles that they reflect very little light, making its thickness virtually unnoticeable.

A large moon may orbit its planet in a predictable way, but the particles in Saturn's rings are engaged in very complicated motions. They are orbiting under the influence of gravity, just as a moon does, but they also collide among themselves.They are pulled hither and yon by the gravity of small moons that are within some rings and large moons orbiting farther out from Saturn, beyond the ring system. And they are subject to the impacts of microscopic and larger meteoroids coming from elsewhere in interplanetary space.

As we discussed at the beginning of the chapter, every 15 years, Saturn is tilted just the right way so that we on Earth are in Saturn's equatorial plane (if you imagine it extending indefinitely far into space). Then the rings vanish from view in all but the most powerful telescopes, because we are looking at them edge-on, and they are extremely thin. Galileo had the same problem, but he didn't know why the objects he spotted alongside Saturn (actually the rings) vanished. But this is the best time to look for very small moons of Saturn (including moons far outside the rings), because it is the only time the moons are not obscured by the rings or lost from view in their glare.

FLYING SAUCER MOONS

We think of the rings as places of mass destruction, their particles coming from the disruption of old moons and from impacts knocking stuff off of current moons or off of larger particles. But things can also grow in the rings of Saturn, at least to a limited extent. Each small moon within the rings has its own Roche lobe, named after the same mathematician as the Roche limit. The Roche lobe is centered on the little moon, and it marks the boundary of the region within which the moon can stick together despite the tidal force of Saturn that pulls upon it. When ring particles collide with the little moon, some of them stick to it and build it up in size, as long as it doesn't extend out past its Roche lobe. If it did, the protuberances would be pulled off by tidal force. Close-up pictures from the Cassini probe show that some small moons are gathering material, not shedding it, under constant bombardment from ring particles.

Two of these little moons, Pan (in the A ring) and Atlas (just outside the same ring), are shaped like flying saucers. Like all the small moons that have been photographed in detail, they have irregular, rough surfaces. At least, that's true for the polar regions of Pan and Atlas. But at the equator, they have smooth bulges that go all the way around the moons. And they are slightly smaller than their mathematically calculated Roche lobes. They began as irregular, rough little moons, probably broken fragments of ancient larger moons that were disrupted. As tiny ring particles collide with them, the particles accumulate in a smooth heap all the way around the midriffs of the moons.

So there are flying saucers out in space, but they are not under the control of little green men. They are formed by natural processes of orbiting systems and colliding particles, all under the action of gravity, tides, and other known forces.

BEYOND SATURN

Past Saturn there are two more gas giant planets, Uranus and Neptune. They are a bit smaller than Jupiter and Saturn, but they have much in common with Jupiter and Saturn, including their own families of many moons. And like Jupiter and Saturn, Uranus and Neptune have rings. Saturn's rings are the most famous because they are the biggest and brightest and can be readily seen in small telescopes like Galileo's. The other three ring systems were not discovered until well into the second half of the twentieth century.

Pluto was called a planet for many years. Currently, it has been designated as a dwarf planet by the International Astronomical Union. Some astronomers still call Pluto a planet, while some don't. Whatever you think about Pluto, keep in mind that it has a moon, Charon, so large in comparison to the (dwarf) planet that many astronomers consider Pluto and Charon a double (dwarf) planet. Pluto also has at least two very small moons, discovered a few years ago with the Hubble Space Telescope.

Pluto is in a region of the solar system mostly beyond the orbit of Neptune called the Kuiper Belt. There are many thousands of small, icy bodies there, called Kuiper Belt objects, or KBOs. Some experts consider Pluto the first known large object in the Kuiper Belt. As astronomers have perfected their methods for searching this distant realm of dim, moving objects, several KBOs have been found that are in the same size range as Pluto, including one that seems to be a bit larger, and are even farther from the Sun. Many of the KBOs, at least the smaller ones, may be comets, meaning they are made almost exclusively of water ice and other frozen substances, with a bit of rock dust mixed in. They don't show comet tails because they are so far from the Sun that there's no heat to warm the ices and turn some of them to vapor that can stream away from the comet and thus constitute a tail.

Even farther out, wherever astronomers look with sufficient instrumental capability, we are finding planets orbiting around other stars. So far, these "exoplanets" are all much larger than the Earth. But we are just beginning to put into operation instruments, telescopes, and space observatories that are capable of finding small planets like the Earth. They will be found before long.

CHAPTER 7

Venus

VENUS IN THE SEVENTEENTH CENTURY

When Galileo first observed the planets in the fall of 1609, Venus was visible only in the early morning sky. Apparently Galileo was not an early riser, since he did not turn his telescope on Venus until the following autumn, when he could see it, conveniently, in the evening. But surely he must have eagerly anticipated what that first glimpse of Venus would reveal. Venus is the mother of the planets: it outshines everything in the sky except for the Sun and the Moon, so brightly, in fact, that on moonless nights it casts faint shadows on the ground. Though you may not have recognized it, you have probably seen Venus many times, for it is often the first object to catch one's eye at night. When it is east of the Sun, we call it the "evening star," stark and brilliant against the indigo twilight, and when it is west of the Sun, it is the "morning star," shining brightly in the gathering glow of dawn. If you know where to look, you can even see it in the bright blue of the daytime sky.

Ancient astronomers venerated Venus, perhaps thinking it drew its brilliance from some inner vital force. She was Ishtar to the

Babylonians, both a mother deity and a goddess of war, and to the Greeks she was Aphrodite, the goddess of love. To academic scholars in the time of Galileo, Venus was one of the seven planets described by Aristotle, presumably traveling through the heavens somewhere between Mercury and the Sun.

Venus is distinguished not only by its brightness, but also by the fact that it always appears close to the Sun in the sky. Depending on the year and the season, Jupiter and Saturn can be found anywhere along the ecliptic (the Sun's path in the sky).They can be seen near the Sun at dusk and dawn, or opposite the Sun, high in the sky at midnight. Venus, however, is never visible at midnight. It is always within about 45 degrees of the Sun and so is only seen within a few hours of sunset or sunrise. As the Sun moves around the sky during the year, Venus travels sometimes to its east and sometimes to its west, always reversing its direction before it gets too far from the Sun. It is like the puppy of the Sun god, bound to its master by an invisible leash.

Galileo would have learned in school that Venus traveled around the Earth, for Aristotle had written that everything in the heavens traveled in perfect circles with the Earth at the center. But astronomers struggled for centuries to reconcile the actual behavior of Venus with Aristotle's dictum of circular motion. If Venus went around the Earth, why did it reverse its direction so frequently, contriving always to stay close to the Sun? Why didn't its motion carry it freely to the side of the Earth opposite the Sun, at least from time to time?

In the second century AD, almost 500 years after Aristotle, the Alexandrian astronomer Claudius Ptolemy devised a geometrical scheme that seemed to solve the dilemma. According to Ptolemy, Venus didn't exactly orbit around the Earth in a single large circle centered on Earth. Instead its path was a circle smaller than its distance from Earth, called an *epicycle*. The epicycle was centered not on the Earth but on a point on the rim of a larger circle, called a *deferent*, which *was* centered on the Earth. It was the center of the epicycle that moved along the deferent, drifting steadily eastward around the Earth at about the same speed as the Sun, while Venus herself revolved on the epicycle. Ptolemy visualized his scheme in purely mathematical terms, but it resembled the wheel-within-a-wheel of a giant amusement park ride. If the speed of Venus around the epicycle and the speed of the center of the epicycle around the Earth had just the right values to synchronize with the Sun, the combination of motions neatly reproduced the Sun-hugging behavior of the planet.

But Galileo was also aware of an unorthodox view that explained Venus much more simply. In 1543, the Polish clergyman Nicolaus Copernicus had published a book called *De Revolutionibus Orbium Coelestium* (in English: *On the Revolution of the Heavenly Spheres*) in which he proposed that the Sun was the center of the solar system, and the planets Mercury, Venus, Earth, Mars, Jupiter, and Saturn orbited around it. In Copernicus's system, the Sun's apparent motion through the constellations was a result of our seeing the Sun from different vantage points as we orbited around it. And Venus never appeared far from the Sun because its orbit was smaller than the orbit of the Earth. When Venus seemed to reverse direction and move from one side of the Sun to another, we were actually seeing it swinging around in its orbit, first passing between the Earth and the Sun on the near side of its orbit and then rounding the curve and apparently reversing as it moved behind the Sun on the far side of its orbit.

Free-thinking souls like Galileo and the German astronomer Johannes Kepler suspected that Copernicus was right, but his ideas were far out of the mainstream of sixteenth-century science. They flew in the face of common sense, which taught us that the Earth was solid and stationary, not flying pell-mell through space. The Copernican system violated the fundamental tenet that the Earth was in the center of the Universe and that everything revolved around it. It even cast doubts on passages from the Holy Book, such as the Old Testament story about Joshua commanding the Sun to stand still during the siege of Jericho.

It was one thing to consider the Copernican theory as an elegant solution to the mathematical problem of reproducing the observed motion of Venus and the other planets. As such it was just a clever idea, like the Earth-centered circles-upon-circles of the Ptolemaic system. Both systems "saved the appearances," according to medieval astronomers—that is, they both made essentially the same predictions for the way the planets moved, and they could both be regarded simply as mathematical tools for calculating the positions of the planets. If you add a column of numbers from the bottom up or from the top down, you get the same result either way, so who is to say which method is correct?

But to determine which is the *real* way the heavens are constructed, we need convincing evidence that rules out one or the other of the two competing systems. In 1610, there was no such evidence, however. Whether the Earth moved or the Sun moved, it made no difference to how things appeared, at least as far as anyone could tell. The believers in a stationary Earth, in fact, could bring up a bit of missing evidence that pointed in their favor. If the Earth orbited the Sun, then during the course of the year, the apparent positions of the stars should shift back and forth a bit in the sky, an effect called *annual parallax*, as the perspective from the orbiting Earth changed. No one had ever seen such an effect, ergo the Earth wasn't moving. (We know today that the distance to the stars is much larger than anyone supposed in the 1600s, and the annual parallax, though it indeed exists, is so small that it could not be measured for another two centuries.) So Galileo, however much he leaned toward Copernicus, was ever the good scientist. Though he believed that the Copernican theory was too attractive not to be the real way the heavens worked, he needed incontrovertible facts that would prove it true.

Venus provided the facts Galileo was looking for. He began observing it sometime in September of 1610, not long after his first observations of Saturn. Galileo was disappointed at first. For all its brilliance, Venus looked like a featureless blob of light with not a moon to be seen. After a few months of observations of both Venus and Mars, he regretfully concluded, in a letter to a friend, that there was nothing of interest to be seen going around these planets.

But Venus continued to hold his attention, because it seemed, almost imperceptibly at first, to be changing shape. The featureless dot he recalled from September had changed to a somewhat elongated oval by early November, and it seemed to be growing in size, though through the fuzzy optics of Galileo's telescope, it was hard to tell whether this was significant or just a trick of the eyes or of memory. But by mid-November, the trend was clear: it was evident that Venus was growing in size and showing a distinct asymmetry in shape. Shortly before the end of December, it appeared distinctly as a half-disk of light, one side notably circular and the other straight.

If he hadn't thought of it before, Galileo now realized he was onto something big. One of his students, Benedetto Castelli, wrote to him on December 5, 1610, suggesting that Galileo's observations of Venus might provide the evidence needed to prove Ptolemy wrong and Copernicus right. Galileo probably knew this already, and he certainly agreed.

Ptolemy's system of epicycles and deferents required Venus to spend all its time between the Earth and the Sun. We on Earth, then, would always see some of the night side of Venus—the side facing away from the Sun—and so Venus would never appear to be fully lit, like the full Moon. On the other hand, if Venus went around the Sun, as seen from the Earth, it would appear to go through a full cycle of phases, from thin crescent when it was nearest the Earth to fully lit when it was on the opposite side of the Sun from the Earth. Was the apparent change he was seeing in the size and shape of Venus a result of its changing phases and distance as it moved from behind the Sun in the summer of 1610 to a position between the Earth and the Sun in early 1611?

As he had with Saturn, Galileo rushed to establish his priority with an anagram. Over a year had passed since his first demonstration of the telescope, and he knew that he now had competition. There were a few good telescope makers besides him and a growing number of astronomers using them to probe the sky. So he wrote a letter to Giuliano de' Medici, the brother of his patron Cosimo, and the ambassador to Prague, where Kepler lived. At the top he put the scrambled letters announcing his discovery, disguised as a Latin message: "*Haec immatura a me iam frustra leguntur o y*," meaning "These premature from me are at present deceptively gathered together o y" (Galileo had not been able to fit the "o" and the "y" into his cipher, so he just wrote them at the end).

By the "deceptive" gathering Galileo may have been alluding to competing astronomers, but in fact he was the only one who had noticed what was going on with Venus. Kepler, when he received the message, tried bravely to decipher it, as he had with Galileo's anagram on Saturn, but all he could come up with was "*Macula rufa in Jove est gyrator mathem, etc.*," meaning "There is a red spot in Jupiter which rotates mathematically." (Jupiter does have a red spot, but it was not observed until the 1650s, by which time both Kepler and Galileo were dead. Kepler seemed to have a talent for making prescient errors.)

By New Year's Day of 1611, it was clear to Galileo that Venus was taking on the appearance of a crescent moon, the flat side of December's half moon turning into a definite concavity. Galileo was ready to reveal his discovery, and he sent the following key to the anagram to his correspondent in Prague, who he knew would pass it on to Kepler. Rearranged, it read, "*Cynthiae figuras aemulatur mater amorum,*" or in English, "The mother of love emulates the shape of Cynthia." Galileo was describing what he had suspected from his observations over the past three months and what was now distinct and clear through his telescope. The "mother of love"—that is, Venus—was going through a complete cycle of phases, just like the Moon ("Cynthia"). Ptolemy was wrong. Venus was in orbit around the Sun.

If he had been cautious about embracing the Sun-centered theory before this, Galileo was now convinced. Henceforth he would be a whole-hearted champion of the Copernican system, a stand that he would defend eloquently in his writing in years to come and that would eventually get him into trouble with the Church and lead to his house arrest in the waning years of his life.

Galileo's observations of Venus established another important fact about the planets that had not been evident in his observations of Jupiter or Saturn: the planets shine by reflected sunlight. That fact, a necessary consequence of the changing phases of the Moon and Venus, seems so obvious to us today that we hardly give it a second thought. But until the 1600s, there was considerable debate over whether the planets reflected sunlight or shone from their own internal luminescence. Kepler, surely one of the more perceptive and inventive of Galileo's contemporaries, responded in a letter that he was surprised to find that Venus was reflecting the light of the Sun, "for on account of the unusual brightness of Venus, I believed light of its own to be inherent in it." He could not imagine how a nonluminous body could be so brilliant: "Astonishing, unless Venus is all gold, or . . . amber."[10]

Kepler, for all his cleverness, was as far from the mark on this as anyone could be, but the true constitution of the planet Venus remained unclear for another four centuries. Despite being able to distinguish its phases, Galileo's telescope revealed no features at all on the Venusian surface. Improvements in telescope size and construction over the decades revealed little more than a disk of unblemished white. Venus, it soon became clear, is covered by a perpetual veil of clouds. Until the first robotic spacecraft were able to land on its surface in the 1970s, no one knew for certain whether its surface, like its phases, resembled the rocky face of the Moon or whether, as Kepler imagined, it was a planet shining with gold and amber.

VENUS TODAY

Today we know that Venus is a rocky planet that formed as a virtual twin of the Earth, a bit closer to the Sun. Venus is just 396 miles smaller in diameter than Earth. But it has changed over billions of years, and it now differs from Earth to an enormous degree. Venus is much hotter, drier, cloudier, and more volcanic than Earth. Its year is shorter than our year, but its day is much longer than ours. It goes around the Sun in the same direction that Earth and the other planets do, but as it orbits the Sun, it spins on its axis in the opposite direction. If you looked down on the Earth and Venus from far away in space above the North Pole, you would see the Earth turn-

---

[10] Drake, Stillman. "Galileo, Kepler, and the Phases of Venus." *Journal for the History of Astronomy.* Volume 15, 1984. 204.

ing counterclockwise while Venus turns clockwise. They are both moving counterclockwise around the Sun.

Venus takes longer to make one turn on its axis (243 days) than it does to make one orbit around the Sun (225 days). On Earth, of course, one turn on the axis is 24 hours, the length of the day, and one orbit around the Sun, our year, is 365¼ days. One day on Venus, judged from sunrise to the next sunrise, is 117 of our 24-hour days.

Earth rotates from west to east, making the Sun seem to rise in the east and set in the west. But on Venus, sunrise is in the west, and sunset is in the east. You wouldn't actually see the Sun if you were there because of the perpetual thick cloud cover, but you would know where it was, thanks to a bright patch of light that you would see moving across the cloudy sky all day. Sunrise and sunset are big deals on Venus, because they occur only twice per (Venus) year. Sunrise to sunset is about 59 Earth days.

The Moon doesn't rise in the west or the east, because Venus doesn't have a Moon. It also has no measurable magnetic field. If you were to hike across Venus, you may as well throw your compass away, because it wouldn't tell you which way was north.The lack of a magnetic field on Venus may not seem like a big deal, but as we will see, it may be the key to why the planet is extremely dry.

There's no liquid water or ice on Venus; there's just concentrated sulfuric acid. The sulfuric acid is present only high up in the clouds, and although there is a continual rain of the deadly liquid, the air is so hot that the drops of acid evaporate long before they can reach the ground. The tiny amount of water found on Venus now is just a trace constituent of the atmosphere, but there probably was much more long ago—perhaps even enough to fill oceans.

Earth has a hydrologic, or water, cycle in which rain falls down to the land, runs off in rivers to the sea, and evaporates, returning to clouds, from which it rains again. (Some water is stored in glaciers, but eventually it melts in place or moves down to the sea where it dissolves into the ocean and once again evaporates.) Venus has a sulfur cycle in which the falling drops of sulfuric acid evaporate and break up into sulfurous vapors that rise into the clouds, condense into sulfuric acid droplets, and fall again. The cycle is wholly confined to the atmosphere and involves no seas, lakes, or glaciers of sulfuric acids or other sulfur compounds. Sometimes sulfuric acid particles rise above the clouds and produce a kind of high-altitude smog or haze. Like the smog over a large city on Earth, the sulfuric haze of Venus comes and goes. The European Space Agency's Venus Express spacecraft, orbiting Venus, found that the haze can appear suddenly, spread over much of the planet, and then disappear in just a few (Earth) days.

The ground is even hotter than the air; the temperature at the surface of Venus is 860°F. There is day and night on Venus, but night gives no relief. The atmosphere is so thick and heavy that it holds the heat in—an extreme case of the greenhouse effect—and keeps the temperature fairly consistent all across the planet. Venus is a very bright "morning star" and "evening star" as seen by us on Earth because its perpetual cloud cover reflects 75 percent of incoming sunlight back into space. Venus is more reflective than any other planet in the solar system. You might think this high reflectivity keeps Venus cool, just as a white or silvery sun reflector in the windshield of a parked car keeps it cool. But Venus's thick atmosphere is very efficient at retaining the heat that does get through the clouds from the Sun. There's no sea level on Venus, because there is no sea, but the air pressure at ground level is 92 times the barometric pressure at sea level on Earth. That's equal to the pressure at a depth of over 3,000 feet in the ocean on Earth.

The clouds form in three distinct layers, covering the altitude range from about 30 to 45 miles above the surface. They are much sparser than clouds on Earth, like a fine mist. But it is a mist that is miles thick, so if you were floating in the Venus clouds and looking up, you would see a certain distance into the mist but not all the way through to the sky unless you were near the cloud tops. And from above the clouds, orbiting spacecraft can't see through to the ground, except by means of radar.

Seen in visible light, the cloud cover is featureless, but when NASA's Mariner 10 space probe passed by Venus in 1974, it took pictures in ultraviolet light, which revealed large, dark markings across the cloud deck, often taking the form of Y- and C-shaped features that covered a large region of the planet. The markings are produced by something in the Venus atmosphere that is a strong absorber of the ultraviolet light coming from the Sun. The substance that produces this effect has mystified scientists for more than 30 years and is still referred to as the "unknown ultraviolet absorber." These striking features could be seen circling Venus rapidly, revealing that the air rushes toward the west at 200 miles per hour, making one complete circuit of the planet in four Earth days. The Mariner 10 scientists discovered and measured this rapid motion, or "super-rotation," of the Venus atmosphere by following the dark markings as they moved across the planet, much like Galileo discovered the rotation of the Sun by charting the movement of sunspots across the solar disk.

At the top of the cloud deck over the north and south poles, there are polar vortices, swirling atmospheric features with eyes like the familiar eye in a hurricane. But on Venus, a polar vortex has two eyes.

The air on Earth is mostly nitrogen and oxygen, with modest amounts of water vapor, carbon dioxide, and other gases. In contrast, the atmosphere on Venus is 96.5 percent carbon dioxide and a few percent nitrogen, with just trace amounts of everything else. The traces include oxygen and water vapor and various gases that contain sulfur. Venus stays hot because the carbon dioxide, which holds in heat, maintains the greenhouse effect. Furthermore, in addition to normal carbon dioxide, the Venus atmosphere contains an unusual form of carbon dioxide that is extremely rare on Earth. Normal carbon dioxide is a molecule that consists of one carbon atom and two oxygen atoms. Each oxygen atom contains eight protons and eight neutrons. The unusual carbon dioxide that is found on Venus is different in that one of its oxygen atoms has 8 protons and 10 neutrons. This so-called isotopic carbon dioxide is more effective at holding heat in the atmosphere than the ordinary gas.

Venus probably formed with a great deal of water, just as the Earth did. The Venus water must have evaporated as the temperature kept rising while the greenhouse effect was building up. Water vapor is good at absorbing heat, so the more vapor went into the air, the more the temperature rose. This is what atmospheric scientists call a "runaway" greenhouse effect, because it gets worse and worse. As water vapor ascended, it was exposed to raw sunlight near the top of the atmosphere. The ultraviolet rays in the sunlight broke up the water molecules into their constituent hydrogen and oxygen. Until very recently, astronomers believed that the hydrogen atoms from the breakup of water vapor molecules simply escaped into space. Hydrogen might do this because its atoms are the lightest of all gases. The oxygen freed from water vapor then reacted chemically with the crust (just as oxygen on Earth reacts with iron to form rust, or iron oxide). So the oxygen was absorbed into the surface of the planet. At least, that's what we thought and what some experts may still think. However, in 2007, the Venus Express space probe discovered that the solar wind (an outflow of electrified and magnetized gas from the Sun that streams outward through the solar system) is sweeping away both hydrogen and oxygen from the top of the Venus atmosphere. Because these gas atoms lose electrons and become electrified, they are susceptible to being carried away by the solar wind. This doesn't happen on Earth to any meaningful degree because the Earth's magnetic field shields us from the solar wind. But Venus does not have a magnetic field, and there is no evidence that it ever did. There may have been oceans on Venus long ago, but they are long gone, and there is not so much as a dry seabed that survives from that era. Earth may have been saved from this fate, retaining the water on which life depends, because our planet does have a magnetic field. Venus was hurt by the lack of a magnetic field and the fact that it is closer to the Sun than Earth, so the solar wind blows past it with greater strength than it does at Earth.

What we see on Venus, rather than ancient sea floors, are almost endless products of volcanic activity: lava plains, volcanoes of various types, and even dry riverbeds. But these river channels are not places where water once flowed. They are structures gouged out of the volcanic plains of Venus by lava that flowed with great speed. Older lava had cooled and solidified, forming the plains, and then new hot lava tore down slopes, creating the channels. The longest "riverbed" on Venus, Baltis Vallis, runs for 4,200 miles, slightly longer than the Nile, Earth's longest river. There are no pyramids along Baltis Vallis, but it does have a typical river delta formation where it ends, like the Mississippi Delta but, of course, dry. Baltis Vallis is also different because it doesn't have tributaries, like the Missouri and Ohio rivers that flow into the Mississippi.

Besides huge shield volcanoes, like Mauna Loa in Hawaii, Venus has mountain ranges that were thrust up by tectonic activity (deformation in the crust of the planet), like the Himalayas. But the Venus mountain ranges are truly unearthly. The highest, called Maxwell Montes, has peaks that reach 38,000 feet above the barren plains of Venus, and a huge, steep slope on one side—toward the southwest—that's unlike anything on Earth.

We have pictures of the ground on Venus at the landing sites of two Soviet space probes. They show dark, flat rocks. If we were there, we would see the rocks bathed in red light, because only the red part of sunlight makes it all the way through the Venus clouds to the surface. Blue, yellow, and orange are all filtered out by the clouds and the atmosphere. Other than the pictures of the two landing sites, the totality of our geographical information about Venus comes from radar scans, because the clouds prevent astronomers from photographing the planet's surface from satellites orbiting Venus (the satellites have to be much higher than the clouds, as otherwise air resistance would bring them crashing to the surface). For the same reason, we can't photograph the Venus surface from telescopes on Earth or telescopes orbiting it like the Hubble.

The radar scans reveal all the geological features described here. They also show a very strange phenomenon. Maxwell Montes and the other tall mountains on Venus are all highly reflective of radar above an altitude of about 13,000 feet. These shiny mountaintops may be strewn with metal-bearing minerals, like iron pyrites ("fool's gold"), or a metallic coating of unknown origin.

The huge shield volcanoes, which were built up as lava flowed down their slopes, make up just a small number of the more than 1,000 volcanoes on Venus. There are many domes, which, as the name implies, are round, convex geologic features. Domes are pushed upward as magma comes up from below and presses against the crust. Venus's domes are like the volcanic domes found on Earth (for example, in Valley of Ten Thousand Smokes in Alaska) but are much larger. Some small Venus domes have flat tops and have been termed "pancake domes." Structures like them are not found on Earth. Other dome-shaped features on Venus, called coronae, have concentric rings of ridges and fractured crust. Hundreds of them are over 100 miles in diameter, and, again, there's nothing like them on Earth.

Some astronomers believe that there are active volcanoes on Venus that still erupt from time to time. But we have not caught one in the act, so we are not sure if Venus belongs in the select company of Earth and Jupiter's moon Io, where active volcanism is known to occur. There is also evidence for lightning on Venus, because a form of low-frequency radio emissions called whistlers, which accompany lightning on Earth, has been observed by instruments on spacecraft near Venus. But no one has caught the actual flash from a bolt of lightning there. If we ever do, Venus will join Earth, Jupiter, and Saturn as a recognized planet with lightning.

Much of the surface of Venus consists of lava plains, mostly lowlands marked with volcanoes, domes, coronae, "rivers" like Baltis Vallis, and other volcanic features. The rest of Venus is marked by strange highlands called tesserae. They are regions of severely disrupted crust that has been folded repeatedly and broken into blocks. The tesserae are older than the lava plains, because the lava plains intrude on their edges as though the lava flowed just that far.

There are impact craters on Venus, too, just as on the Moon, the Earth, and many other bodies in the solar system. Generally, impact craters are the result of large meteoroids or small asteroids smashing into the surface of a planet or moon with great force. But the craters found on Venus are unusual. The meteoroids and asteroids in interplanetary space have a pronounced size distribution: the smaller they are, the more of them there are. Therefore, on the Moon, for example, the bigger craters are fewer than the small craters, because more little objects hit the Moon than big ones. But on Venus, there are more big craters than small ones. At least, there are almost no craters that are smaller than about two miles in diameter. The likely reason is that small in-falling bodies on Venus disintegrate in the thick atmosphere or are decelerated by air resistance to such low speeds that they hit the ground without enough energy to make craters.

On the Earth, erosion by wind and water gradually wears away an impact crater. But the craters on Venus look almost brand new and show little or no sign of erosion. Geologists expected that there would be some large old impact craters that were flooded by lava after they formed but not completely filled in by the lava. But no such lava-overrun impact craters are found in the radar scans. The explanation for this unexpected deficiency may lie in the catastrophic volcanism theory of Venus's surface. According to this widely respected theory, there was a planet-wide spasm of heavy volcanism around a half-billion years ago that completely wiped out all the big craters that formed in Venus's early days. And erosion is so slow on Venus that impact craters that formed since a half-billion years ago still look good as new.

Erosion is weak on the surface of Venus because there's no water and because the wind at the surface is weak. The wind isn't much faster on Venus than a person can walk on Earth. Also, the Venus air is so dense that it impedes any windborne dust particles from striking the rocks with much force. Sandstorms won't pit your windshield on Venus.

There are geologic structures called wrinkle ridges on the lava plains. These ridges are tens of miles long and roughly parallel as though something squeezed and compressed the crust of Venus in their vicinity. All our pictures of wrinkle ridges on Venus come from radar maps, but similar features have been photographed on the Moon and Mars.

Venus has had a hard time, losing its water, getting extremely hot, and being deformed and stretched and squeezed by powerful forces at work on its crust and probably also by an episode of volcanism so great as to completely rework the whole surface of the planet. We're fortunate that Earth has fared so much better.

CHAPTER 8

Comets

COMETS IN THE SEVENTEENTH CENTURY

Between 1610 and 1613, the peak years of Galileo's telescopic activity, no bright comets graced the nighttime sky. By 1618, when not one but three of these celestial attention-getters made unexpected visits, Galileo was no longer a frequent observer of the heavens. Then in his fifties, troubled by physical ailments and by hassles with Church authorities in Rome, he had assumed the role of a graying eminence, devoting himself more and more to his scientific correspondence and to writing books on astronomy and physics for the general public. Though he had little to contribute as far as telescopic observations, the comets of 1618 sparked a sharp public debate in which Galileo got an opportunity to float his own novel ideas about what caused their arresting appearance and erratic behavior.

Galileo knew that comets are always greeted with a mixture of fear and fascination. Unlike the planets, which move predictably around the sky, comets appear suddenly and without warning. They materialize at first as fuzzy blobs of light and then grow dramatically, gradually unfurling long, pearly tails that in some cases stretch from horizon to horizon. Inevitably, they fade and shrink after a few weeks or a few months, disappearing as quickly and surprisingly as they came into being. Comets are also protean: their shapes can change from night to night, and no two comets look or behave identically. Some have fat, fan-shaped tails; others trail long, wispy banners. Some drift slowly and majestically over the span of several weeks; others race through an entire constellation in the course of a single evening. In contrast to planets, which move generally west to east and stick close to the annual path of the Sun (the ecliptic), comets can appear anywhere in the sky and move in any direction.

A bright comet appears about every 10 years, so an average person may see several in a lifetime. Still, comets are rare enough to cause puzzlement and consternation whenever they blossom into view. The Roman philosopher Lucius Seneca, whose writings were quite familiar to Galileo, wrote about the invariable public reaction to the appearance of a comet: "everybody is eager to know what it is. Blind to all the other celestial bodies, each asks about the newcomer; one is not quite sure whether to admire it or fear it. Persons there are who seek to inspire terror by forecasting its grave import. And so people keep asking and wishing to know whether it is a portent or a star."[11]

The consensus was, more often than not, that comets were portents, foretelling famine, pestilence, or war. Or worse, according to the sixteenth-century Lutheran bishop Andreas Celichius, they were "the thick smoke of human sins, rising every day, every hour, every moment full of stench and horror, before the face of God, and becoming gradually so thick as to form a comet, with curled and plaited tresses, which at last is kindled by the hot and fiery anger of the Supreme Heavenly Judge."[12] More often than not, circumstances would bear out the doomsayers—there seems always to be enough famine, pestilence, or war to satisfy any pessimistic prediction, no matter what its source.

In Galileo's time, these superstitions about comets were widely held (they are not uncommon even today), but there was also a more

---

[11] Yeomans, Donald. *Comets*. New York: John Wiley, 1991. 2.
[12] Ibid., 22.

sanguine view of the physical origin of comets that coexisted alongside the myths. It was drawn, as was so much else, from the writings of Aristotle and his commentators. In the Aristotelian Universe, comets were not astronomical objects but meteorological phenomena. Everything in the realm of the planets—from the Moon on out to the stars—was supposedly changeless and eternal, traveling around the Earth in perfect circles. Therefore, reasoned Aristotle, comets could not be out there with the planets. Aristotle believed instead that comets were like swamp gas—vapors drawn out of the Earth by the heat of sunlight. Rising up into the atmosphere, they would be ignited by friction and then carried around by the rotation of the heavenly spheres above the atmosphere. The different shapes and sizes of comets were as natural as the different shapes and sizes of clouds, because both clouds and comets were gaseous accumulations that were molded by winds, light, and other conditions in the atmosphere. The association between comets and famine, in this picture, might even be attributable to especially intense periods of sunlight, which would not only evaporate more gas from the Earth but could also cause lakes to dry up and rains to be delayed.

Aristotle was regarded as the font of all knowledge, the philosopher's philosopher, and his writings had enormous influence throughout the Western world for centuries after his death. There were a few unorthodox views advanced from time to time (Seneca, for one, believed that comets were celestial bodies like the planets), but they never took hold. The notion that comets were meteorological phenomena became the accepted wisdom, taught in school and written about by philosophers, and that was that. But, of course, all of this hung on the authority of Aristotle, and his views about comets, for the most part, were based on speculation rather than hard data, since no one had ever caught the process of comet formation in the act. Even more to the point, no one had ever measured the distance of a comet to check whether or not it was inside the Earth's atmosphere.

The first to actually carry out a reliable measurement of the distance to a comet was the Danish astronomer Tycho Brahe, three decades before the introduction of the telescope. In November 1577, a bright comet appeared in the evening sky, and Brahe, who was already one of the most accomplished celestial observers in Europe, observed it regularly until it disappeared two and a half months later. Brahe was a master of pre-telescopic astronomy. Using instruments of his own design, which resembled large gun sights and circular arcs with carefully graduated angular scales, he was able to measure the separations of objects in the sky better than anyone previously had, achieving precisions of a fraction of a degree.

Brahe's instruments, for the first time, made it possible to measure the *parallax*—the shift in position—of the new comet during the course of a night. Astronomers had known for thousands of years that objects in the sky appear to shift position when viewed from two different places, and the closer the object, the bigger the shift. The easiest way to observe a celestial object from two positions was simply to wait. Since the heavens turn (or, in modern terms, since the Earth rotates), an observer in one spot sees objects in the heavens from a continually changing perspective. The amount of parallax shift for the Moon, when viewed on opposite horizons (i.e., comparing its position when it is rising with its position when it is setting), is about 2 degrees.

Brahe's instruments were easily capable of measuring a parallax of 2 degrees, but he found that the comet barely seemed to change position with respect to the background stars at all. Such a small parallax indicated that its distance was far beyond that of the Moon, clearly in the heavens, and definitely not in the Earth's atmosphere. Brahe concluded that the comet orbited the Sun permanently in a circle a little beyond the orbit of Venus, and that something had caused it to brighten in November 1577 and fade in January 1578. (He may not have been right about the orbit, but he was correct in finding that the comet was beyond the atmosphere.)

Brahe's measurements of the comet of 1577 convinced him that the heavens were not changeless, reinforcing a conclusion he had already reached five years earlier. In 1572, Brahe had measured the parallax of a new star, or *nova* (we now recognize that this was huge star exploding at the end of its life and refer to it as a supernova). He found that the nova also was far beyond the Moon. Yet he was not ready to totally abandon Aristotle's picture of the Universe. In Brahe's book on the comet, published in 1588, he presented his own model of the Universe, a sort of hybrid scheme halfway between those of Aristotle and Copernicus, in which the Earth remained motionless at the center but all the planets (among which he did not count the Earth) orbited the Sun. This concept was never widely adopted, but Galileo surely knew about it as an alternative to the Copernican theory.

By the early 1600s, many astronomers shared Brahe's conclusion that comets were heavenly bodies, a position in agreement with our modern understanding. There were a few notable exceptions: Johannes Kepler believed that comets were atmospheric phenomena that moved across the sky in straight lines like rockets. And when Galileo finally got around to writing about comets, he ended up, like Kepler, on the wrong side of the argument.

The three comets of 1618 that provoked Galileo's comments appeared in rapid succession during the fall of that year: the first in October, the second in mid-November, and the third, by far the brightest, at the end of November. Though he was confined to bed and could not observe, there were by this time numerous astronomers who were using telescopes. However, they were able to see little more than a spread-out image of the diffuse glow visible to the naked eye. The sad conclusion was that comets, unlike the Moon and planets, revealed no distinguishing features through a telescope.

At this time, a friend wrote to the ailing Galileo with news from Rome, telling him that the Jesuits were talking about the new comets and using them to argue against the Copernican theory. The anti-Copernican argument was rather far-fetched in retrospect: the comets showed no parallax, placing them up in the heavens, just as Tycho Brahe had claimed 40 years earlier, so if Brahe was right, Copernicus was wrong (even though he wrote essentially nothing about comets). In a public lecture, and a year later in a published treatise, the Jesuit Horatio Grassi not only cited parallax, but also telescopic evidence. Comets, he claimed, looked no larger through the telescope than they did to the naked eye. Since they were not magnified, they must be farther away, because the telescope magnifies nearby objects but will not magnify distant objects.

From a modern perspective, Grassi had a strong argument. The central dot of light that often appears in the center of a comet's head is unresolved, just as stars are, even through a telescope. No matter how much it is magnified, it still looks like a dot. The reason that stars are unresolved by a telescope even though they are huge objects is that stars are very far away. But the dot of light seen in the head of a comet through small telescopes, which may be the nucleus (the solid body of the comet), or the "false nucleus," a small cloud of icy particles around the nucleus, is just too small to resolve as a disk in small telescopes. The dot is indistinguishable from a star in small telescopes. Grassi's argument was clever, but he went on to draw the wrong conclusion that the comet of 1618 was in a circular orbit around the Sun at about the same distance as Venus. We know today that most comets are in long, looping orbits that take them out to the edges of the solar system. Still, Grassi was correct on the point that comets were celestial, not terrestrial, in origin.

To Galileo, on the other hand, everything Grassi claimed was not only questionable, but outrageous. As he read it, Grassi was saying that telescopes magnify nearer objects more strongly than distant ones when clearly they magnify without discrimination as to distance. Grassi's obtuseness irritated Galileo no end; he was not one to suffer fools graciously. It also rankled him to realize that the Jesuits were misusing the 1618 comet to argue against Copernicus, whose ideas Galileo considered unquestionably established. Had he not proven the validity of the Copernican system by his own well-publicized observations of the phases of Venus?

And so, partly out of sheer contrariness, Galileo was quick to respond, taking an opposing view that today seems even more unconvincing than Grassi's argument. He first attacked Grassi through two lectures that were delivered by a friend, Mario Guiducci, and published in 1619. Grassi replied (under the pseudonym Lothario Sarsi) to Galileo's attack with a small book with the delightfully suggestive title *The Astronomical and Philosophical Balance, on which the opinions of Galileo Galilei regarding Comets are weighed, as well as those presented in the Florentine Academy by Mario Guiduccio.* Galileo's riposte, *The Assayer,* printed four years later, was a major work not just on comets, but on the nature of the scientific method.

Galileo's arguments consisted mostly of sniping at things he didn't like about the anti-Copernican manifestos on comets. He took Grassi to task for making unsubstantiated claims and drawing hasty conclusions. The argument that telescopes do not magnify very distant objects was just plain wrong, he asserted. And a lack of parallax, argued Galileo, does not necessarily mean that something is very far away. For example, rainbows are clearly atmospheric phenomena but do not appear to shift as one moves. Could comets not be atmospheric phenomena of the same sort? Perhaps their tails are produced (as Kepler believed) by some sort of optical phenomenon like a ray of sunlight refracting through a glass of water.

Galileo presented no unified view of comets in this debate, and we need not dwell on all the claims and counterclaims that flew back and forth between him and his adversaries. To take a charitable view, we can look back on Galileo's writings at this time as sensible critiques of weak scientific argumentation, even though Grassi, in claiming that comets orbited the Sun, was more right than wrong. Galileo, if he didn't declare outright that comets were atmospheric phenomena, seemed to be arguing that there was no conclusive proof that they weren't. More research was needed if a secure conclusion was to be reached.

Looking back, then, after four centuries of scientific research, it's hard to fault Galileo's conservatism, even if it seems out of step with his bold prescience in other astronomical matters. If Galileo's views on comets seem hopelessly wrong, they tells us less about his scientific insight than about how little hard evidence there was in those days to support any firm conclusions about the nature of the heavens. The telescopic age was just beginning, and despite the remarkable advances Galileo had made with his observations of the Moon and the planets, so much remained to be discovered. We don't doubt that Galileo, for all his self-assurance, would have agreed with Seneca's comment made 1,500 years earlier: "The day will yet come when posterity will be amazed that we remained ignorant of things that will to them seem so plain. . . . Men will some day be able to demonstrate in what regions comets have their paths, why their course is so far removed from the other stars, what is their size and constitution. Let us be satisfied with what we have discovered, and leave a little truth for our descendants to find out."[13]

COMETS TODAY

Today we know that comets are real objects and not optical illusions as Galileo had thought. But they remain among the least understood bodies in the solar system, despite all that astronomers have learned from observation with modern telescopes and from visits to comets by instrumented space probes.

It's hard to draw firm conclusions about comets, because they come and go, they seem to differ from one to another, and their orbits around the Sun can be dramatically changed because of gravitational effects from larger bodies, such as the planets or even passing stars on the distant outskirts of the solar system. So when we see a comet in a certain orbit, we may not know how it got there, where it came from, or how long it's been in that orbit. But in some cases, we see the effects of, say, Jupiter's gravity as it redirects or even disrupts a comet.

The main body of a comet, the only solid, more or less permanent part, is the nucleus. It's a big lump of frozen water and other frozen substances, collectively called "ices," interspersed with small rock particles, or "dust" in astronomers' lingo. For that reason, the nucleus is sometimes called a "dirty iceball," a term that implies that the ices make up the bulk of the object. However, studies of one cometary nucleus indicated that it has more dust than ice by mass, implying in this case that a better term would be "icy dirtball." It's likely that both kinds of object exist—iceballs and dirtballs.

The comet nucleus is typically a few miles in diameter, and if the comet is in a small enough orbit around the Sun so that it sweeps through the region of the planets, around the Sun, and back out

---

[13] Ibid., 10.

into space every so often, the nucleus is shrinking. The ices sublime, meaning that as they are warmed by sunlight, they turn directly from the solid state to gas without ever becoming liquid. This process is what you see when you have a block of dry ice (frozen carbon dioxide) at room temperature: vapor comes off of the dry ice, and the block shrinks before your eyes, but it never melts. When a block of dry ice gives off vapor, the vapor is carbon dioxide gas. Most of the gas that comes off of a comet is water vapor, although there are many other substances as well. We are used to thinking of water vapor as steam. But the water vapor from a comet is very cold. The gas produced when the Sun warms the comet nucleus streams off into space, blowing dust away from the nucleus as well. So the nucleus gets smaller each time the comet rounds the Sun, typically losing a small fraction of 1 percent of its mass, an amount of matter equivalent to a layer of surface material about 1 yard thick. This number, like all the numbers we use to describe comets, can be very different from one comet to another, and in most or even all cases, the numbers are just rough estimates.

The comet nucleus is usually not round like a planet or a basketball. It is likely to have a pronounced long axis, lumpy hills, pits or craters, and other surface irregularities. Its shape is more like an Idaho potato than anything else. And it looks like a potato that has been baked in the embers from a campfire: it has a black skin, called the crust. But this crust has very little in common with the crust of the Earth. The crust of a comet nucleus is so dark that it reflects only a few percent of the sunlight that falls on it. The nucleus crust is hotter than a lump of pure ice would be at the comet's distance from the Sun, because the dark material absorbs sunlight and heats up, just as dark-colored cars parked in the sun get hotter than light-colored ones.

Suppose you took a hot baked potato and jabbed it in a few spots with a knitting needle. You would see some steam emerge from the fresh holes. The comet nucleus behaves much the same way, but the vapors that emerge are cold. The cold gases don't come off the whole surface of the nucleus; they come mostly, if not entirely, through a few pits in the crust. That's where the analogy with a potato comes to an end. The pits only blow off vapor when the Sun shines on them. The nucleus is rotating, just as the Earth and other celestial bodies turn, and the pits erupt while they are facing the Sun. Then, as they rotate out of the sunlight, they turn off, and then they turn on again as they rotate to face the Sun again. The Earth takes 24 hours to make one turn on its axis, but a comet nucleus, although much smaller, may take longer to rotate. The nucleus of Comet Halley was turning about once a week when it was visited by three spacecraft in March 1986.

Several comet nuclei have been examined closely by space probes, and each one has differed to some extent from the others. Some show evidence that they may have a layered structure, like the strata that are exposed in the walls of the Grand Canyon but much less pronounced. If there are indeed layers in a comet nucleus, that might indicate something about how the material was deposited when the comet formed. However, other observations reveal that the gas coming off of the nucleus of a particular comet can differ in chemical composition from one day to another, as though the nucleus formed by the merging together of separate clumps of icy material, which have been dubbed cometesimals.

In close-up photos from big telescopes and spacecraft, the gas erupting from a pit in the comet nucleus looks like a bright jet of material. The vapor itself may be invisible, or nearly so, but it carries dust particles and probably little grains of ice, and the dust and ice grains reflect sunlight. Other kinds of pits that appear to be impact craters have been found on one comet nucleus, and steep-sided promontories and spires several hundred feet tall have been found as well.

The jets of escaping matter from a comet nucleus act like little rockets, pushing back on the nucleus. The basic motion of a comet around the Sun is just like Earth's—that is, it's a result of the gravitational attraction of the Sun. There are also effects from the gravity of planets that the comets pass, since they often travel in long, elliptical orbits that cross the more nearly circular orbits of many of the planets. In addition, however, the jets propel the nucleus out of its normal orbital motion enough that astronomers detected the changes in comet orbits long before the jets themselves were discovered. These changes in the orbit of a comet are called "nongravitational effects."

Although we know the typical comet nucleus is anywhere from a few hundred yards wide to a few dozen miles across, very little is known about the total mass of the object or its density. It must be less dense than rock, but it's not necessarily the density of its main constituent, water ice. It may be relatively porous, for example, and therefore could be less dense than ice.

We don't know much about what's inside a comet nucleus. We do know there's water ice and other frozen gases in lesser quantities, as well as rock dust, and the interior well below the crust is very cold—probably -370°F or less. If the nucleus is made up of stuck-together cometesimals, as previously suggested, it would explain why comets occasionally shed a few pieces or even break up into a lot of pieces of about the same size, like bricks that tumble down from an old wall.

The dark crust of the comet nucleus is probably composed of carbon-bearing substances. Organic compounds have been detected in comets and could be part of the dark material. There might be rock particles in the crust, too, that would include minerals such as olivine and pyroxene. According to comet dust samples collected from Comet Wild 2 and returned to the Earth by the Stardust spacecraft, the microscopic particles are little rocks—that is, they are not individual minerals but stuck-together assemblies of minerals, like common Earth rocks, in which you can see different mineral components jumbled together.

Some comets that are approaching the Sun from a great distance appear to brighten suddenly and for a brief period while they are still very far from the Sun. This probably occurs when a large part of the crust is blown off the comet because of a vapor buildup inside the nucleus. The dust in the erupted crust spreads out into a cloud that reflects more sunlight than the nucleus itself, so the comet is briefly much brighter than before. As the dust dissipates, the comet returns to its normal brightness for its current distance from the Sun.

The cause of the crust eruptions in certain incoming comets may be related to the type of water ice in the nucleus (although there are other theories). Ordinary ice on Earth, like the ice cubes in your freezer or the ice on a pond in winter, is crystalline, like a diamond. The crystals may not be the same shape, or as hard and heat-resistant, as diamond crystals, but they are crystals nonetheless. In the nucleus of a comet, however, water ice is very likely in amorphous form, meaning that the atoms and molecules are not systematically arranged in a repeating pattern as they would be in a crystal, but are bunched together with no apparent structure. There's no amorphous ice in nature on Earth, because it forms only at extremely low temperatures, which don't occur on Earth. But the interior of the comet nucleus is much colder. It probably formed largely from amorphous ice and then remained very cold. Amorphous ice will sometimes change to crystalline ice in a phase change, much like crystalline ice can change to liquid water or, under space conditions, directly to a gas, water vapor. This can occur when the Sun warms the crust of a comet and the heat is conducted into the ice layers below. When amorphous ice changes to crystalline ice, it releases heat. It's that heat that is believed to make some of the other frozen substances in the nucleus turn to gas, and the expansion of that gas then lifts off some of the comet crust, producing a temporary cloud of dust that reflects sunlight and makes the comet much brighter.

Most of the above information about comets is different from what most people think or know about comets. That's because— except for the historical account centering on Galileo's thoughts— we have yet to mention the head or tail of the comet, the parts that most people know about, which are seen most easily in a photo of a comet or by eye when a bright comet appears in the night sky.

The "head" of a comet is a nontechnical term that refers to the nucleus plus the much larger and more spectacular *coma*, a cloud of gas and dust that surrounds the nucleus as it gets close to the Sun, especially when it's inside the orbit of Mars. The coma consists of matter that has blown off the nucleus through the jets in the crust. Besides the coma that we can see, there is a much larger coma of hydrogen gas, called the hydrogen cloud or hydrogen coma, which is invisible even through telescopes but can be imaged in ultraviolet light by instruments in space. (Ultraviolet light is blocked by the Earth's atmosphere, so it can't be received by telescopes on the ground.)

Gas in the coma is bathed in the harsh rays of the Sun, especially solar ultraviolet rays, which cause gas molecules to break up into atoms. The ultraviolet light also causes molecules and atoms to lose one or more of their electrons. When an atom or molecule is missing an electron or two, it becomes an electrically charged ion. It is then subject to being blown away from the comet and farther from the Sun by the solar wind. Those atoms or molecules that do not lose electrons (they are called "neutral gas") are also pushed away from the comet and from the Sun by the pressure of sunlight.

There's dust in the coma, too, which comes out of the nucleus through the jets. The dust particles are blown away from the comet by the pressure of sunlight, just like the neutral gas. Examination of dust particles from Comet Wild 2, brought back by the Stardust mission, shows that most were formed in the early solar system—not in interstellar space—before the Sun and planets began to form. At least one of these comet particles formed at a high temperature close to the newborn Sun and somehow made its way to the outskirts of the planetary region where the nucleus of Comet Wild 2 was born, incorporating the particle.

The frozen chemical compounds in the nucleus are called parent molecules. As the parent molecules are broken up in the coma, they yield individual atoms and smaller molecules. Besides the microscopic dust, there are larger particles of frozen material, which may be a half-inch or even an inch in size, that come off the nucleus in large numbers and sometimes form a recognizable inner coma. Each of these large particles or ice grains sublimes and lets off gas and possibly dust, too, just like the nucleus, but before long the grain is all gone, like a snowflake that melts away on a warm window.

The dust particles that are blown out of the coma form a long, smooth, curved structure in space behind the comet. It's called the dust tail and is usually the bright tail that we see on a comet. When the comet is bright enough and the sky is dark enough, we can see that the dust tail has a yellow tint, because we are viewing it by means of the sunlight that reflects from the dust.

Photographs of comets show that a typical dust tail can extend to great distances from the nucleus. However, spacecraft have revealed that far beyond the visible dust tail, an unseen dust train persists over even vaster distances. This dust train is also called a meteor stream, because when the Earth passes through the train or stream, as it does in the case of certain comets, large numbers of the dust particles strike the atmosphere. When a dust particle of sufficient size penetrates down through the upper atmosphere on the night hemisphere of the Earth, it lights up the air molecules along its path, and people below see a meteor, or so-called "shooting star." When there are a lot of meteors, all coming from roughly the same direction, they form a meteor shower, like the famous Perseid meteors that flash across the sky every August. And in some cases, astronomers have compared the orbits of the meteor particles with the orbits of known comets and have determined which comet shed the dust train that causes that meteor shower every year. The Perseid meteor shower, for example, is caused by dust shed from Comet Swift-Tuttle, which was discovered in 1862 and which follows a very elongated elliptical orbit around the Sun, taking about 133 years for each complete circuit. Although Comet Swift-Tuttle appears at most once in a person's lifetime, the dust that it has shed over many years is spread all around the Sun in a great elliptical structure, like an out-of-round donut, and the Earth passes through it every August.

Some comets have another visible tail that is bluish, straighter than the dust tail, and not smooth at all but composed of knots and filaments. This is the ion tail, consisting of ions pushed away from the coma by the solar wind. Its electrified particles move along lines of the magnetic field from the solar wind that have wrapped around the coma in a clothespin, or long U, shape, with the bottom of the U pointing toward the Sun and the legs of the U pointing opposite the Sun, extending for tens of millions of miles. The ion tail is like a windsock, showing pilots and air traffic controllers which way the wind is blowing at an airport. However, this tail is a windsock that tells astronomers which way the solar wind is blowing at the comet's location in the solar system.

The ion tail can have individual straight "rays" or crooked linear structures, and sometimes astronomers see a bright knot of ions with a higher density than those in the rest of the tail. The knot can be tracked as it zips outward along the tail. It moves faster and faster until it reaches the speed of the solar wind. (The wind speed varies depending on position in the solar system, but it is about 1 million miles per hour at the Earth's distance from the Sun.)

Ion tails have a bluish tint because the ions don't reflect much sunlight but produce their own light. In particular, ions formed from the carbon monoxide molecule when it loses an electron give off the blue light that colors the ion tail.

As the comet moves through space in its orbit around the Sun, it sometimes crosses a sector boundary in the solar wind. A sector boundary is a place where the magnetic field in the wind goes from pointing one way to pointing in the opposite direction. When the ion tail, which is a magnetized structure itself, crosses a sector boundary, there's a kind of magnetic short circuit. This disconnects the ion tail from the comet, and the tail flies away. Soon, the comet grows another ion tail, like a lizard that is recovering from a close call with a predator. All the material in an ion tail (as well as in a dust tail and the coma) is ultimately leaving the comet (meaning the nucleus) and dissipating into space. However, when an ion tail breaks off, it flies off at a greater speed.

WHERE COMETS COME FROM

Comets formed in the earliest days of the solar system, about 4.6 billion years ago, on the outskirts of the planetary system, probably beyond the orbit of Saturn, where cold matter from the solar nebula, the birth cloud of the Sun and planets, merged to form icy bodies that could not have survived long close to the Sun. They formed in a large flattened volume of space centered on the Sun, just as the planets formed in roughly the same plane, which passes through the equator of the Sun. Many large comets and related icy bodies still persist in this region, some outside the orbit of Neptune in an area called the Kuiper Belt, after Gerard Kuiper, one of the astronomers who deduced its existence.

Very few of the hundreds of *observed* comets, however, are in the Kuiper Belt. And there may be huge numbers of comets out there that we don't have the technical capability to detect, although that technology is coming. Here's how most of the known comets are categorized:

* long-period comets
* dynamically new comets
* sun-grazing comets
* short-period comets
* Jupiter-family comets
* Halley-type comets
* main-belt comets

We need to emphasize that the following descriptions of these comets' origins are uncertain. They're the best that astronomers can do based on computer experiments called numerical simulations. If you read another book on this topic you probably will find a slightly different story, because experts keep making new and hopefully better simulations with more powerful computers and sometimes even better input data, as they become available.

Astronomers believe that there are about a trillion comets in the region around 20,000 Astronomical Units, extending out to about 75,000 A.U., from the sun, where they move through a huge, nearly spherical zone called the Oort Cloud. It's named for Jan Oort, the Dutch astronomer who deduced its existence. They circle the Sun in long slow orbits that never bring them within view, unless they are disturbed by a passing star or some other gravitational effect extending from beyond the solar system. When that happens, some comets head in toward the Sun where we may discover them, and others, which we have never seen and never will see, are thrown on escape trajectories and leave the Oort Cloud (considered the outermost reaches of the solar system) forever. Some of the comets that head inward are discovered and classified as dynamically new. Passing the planets, they are often further perturbed by planetary gravity so that the outermost point of their orbit, called aphelion (meaning furthest from the Sun), moves far inward. If this occurs, they will come around the Sun again and again until they dissipate from losing mass again and again into their coma and tails. (Out in the Oort Cloud, the Sun's light is so feeble that comets are probably just bare nuclei, with no meaningful coma or tail. They are ice balls in cold storage.)

The comets in the Oort Cloud are moving every which way; they are not confined to a flattened system around the plane of the Sun's equator, like the planets and asteroids and certain other comets. Some orbit the Sun in the same manner as the planets, moving counterclockwise as seen from above the North Pole of the Earth. But about as many are orbiting in the opposite, or retrograde, direction. Their orbits collectively trace out a great sphere around the Sun.

A comet that comes in from the Oort Cloud and is captured into a smaller orbit that brings it repeatedly within the orbits of at least the outer planets joins the class of long-period comets, defined as those that take more than 200 years for each pass around the Sun. At least, that's the leading theory for where long-period comets come from.

Dynamically new comets are long-period comets that are seen as they make what is believed to be their first-ever pass into the planetary system, crossing the orbits of several, many, or all of the planets. The inbound legs of their orbits indicate that they are coming from a region located around 20,000 times the distance between the Earth and the Sun. By contrast, Pluto's average distance from the Sun is just 39.5 times that distance. It is the dynamically new comets, on their first approach toward the Sun, that often brighten dramatically, probably as their crusts erupt.

Sun-grazing comets are generally long-period comets that come very close to the Sun. Several hundred of them have been seen, most of them very small and only detected with satellite instruments that make it possible to see objects very close to the Sun. When these small comets are far from the Sun, they are so dim that very few of them can be seen. Near the Sun, they reflect so much sunlight that they are detectable. Many have orbits so similar that they are thought to have arisen from the breakup of a single very large comet. On rare occasions, a large sun-grazing comet comes by, like Comet Ikeya-Seki in 1965. It became so bright as it approached the Sun that on the day of closest approach, it could be seen next to the Sun with the naked eye.

Comet experts suspect that that there is an inner Oort Cloud of comets at a distance of roughly a few thousand astronomical units from the Sun. By some estimates, it has five times or more comets than the Oort Cloud proper, but other computer simulations suggest that the two "clouds" have about the same number. The comets in the outer Oort Cloud are so far from the Sun and so weakly attracted to it that many will be pulled from the cloud by the gravity of passing stars, which can sometimes go right through the cloud, or by gravitational effects from matter in the Milky Way Galaxy surrounding the solar system, or even by rare approaches of huge, cool clouds of molecular gas in the galaxy. In the past, these effects may even have emptied the Oort Cloud over billions of years. It may then have been refilled with comets escaping from the inner Oort Cloud.

Short-period comets take less than 200 years to orbit the Sun. Their orbits are not randomly oriented like those of the Oort Cloud comets but are predominantly located in a fairly flattened system, like the orbits of the planets but not as flat. At least, that's true for those with orbital periods of less than 20 years, which are called the Jupiter-family comets. The Jupiter-family comets go around the Sun in the same direction ("prograde") that the Earth does. This hints at their origin. They probably come from the Kuiper Belt and are moved inward by gravitational effects from the outer planets. The short-period comets with orbital periods of more than 20 years include Halley's comet and are sometimes called the Halley-type comets. Their orbits are more random than those of the Jupiter-family comets, and in fact, Halley's comet itself orbits in the retrograde direction.

The main-belt comets are a small class of comets (at least, only a few have been found so far) that orbit the Sun in the asteroid belt that is located between the orbits of Mars and Jupiter. ("Main belt" is the technical term for the asteroid belt, as there are some asteroids in other regions of the solar system.) They may have formed in a different region than other comets—perhaps even in the main belt, although how they could form with a significant amount of ice that close to the Sun is not clear. Not much is known about them so far.

THE FATE OF COMETS

Some comets simply shrink and perhaps dissipate into space as a result of repeated swings past the Sun. That's what is happening to Comet Halley, which comes by about every 76 years. Others break up into smaller chunks that then dissipate at a faster rate than the larger nucleus that spawned them. Some pass Jupiter or another large body too closely and are torn apart by tidal force. That happened to Comet Shoemaker-Levy 9 when it passed very close to Jupiter in 1992. It was broken into about 20 large pieces, which each grew a tail, and the pieces ultimately smashed into Jupiter one after another in July 1994. Other comets just seem to simmer down; their jets stop spouting, perhaps because a thicker crust has built up, or because many of the volatile ices in the outer layers of the nucleus have been exhausted, or both. At that point, these so-called dormant or extinct comets resemble asteroids, the small rocky bodies that orbit the Sun mostly in the asteroid belt. But the dormant or extinct comets tend to move in more elliptical orbits than the average asteroid—one hint that they once were comets. And finally, some comets remain far out in the Oort Cloud, or the inner Oort Cloud (if it exists) forever, staying cold and never growing a coma or tail.

The Shoemaker-Levy comet fragments that smashed into Jupiter were dramatic evidence that comets do strike planets. It is believed that early in the history of the solar system, there were large comet showers in which many comets hit each of the planets, including Earth. Some astronomers believe that the impacting comets contributed measurably to the amount of water in the oceans. Chemical analysis of the relative amount of hydrogen isotopes in Earth's water and in comets might provide evidence for or against this theory, but the data are not good enough yet for a firm conclusion. And it's possible that some comets formed with a different chemical and isotopic makeup than others, so many comets must be accurately studied before we will know if the theory is true. Furthermore, we know that there is much organic material—carbon-bearing molecules that may be building blocks for life—in comets. So comets that smashed onto the early Earth may have provided raw material for life. Maybe, or maybe not.

CHAPTER 9

The Stars and the Milky Way

THE STARS AND THE MILKY WAY IN THE SEVENTEENTH CENTURY

These days, unless you live in the remote mountains of Montana or the Australian outback, it's hard to appreciate how dark the skies must have been in Galileo's time. No bright streetlights competed with the faint glimmer of starshine, no store marquees distracted nighttime strollers, and no security lights flooded the sky with their glare. Whether you were in urban Venice or in an isolated rural village, you could look up on a clear moonless night and see utter blackness punctuated by a vast number of stars. This appearance of stars at night, without a doubt, was the very vision of infinity to people in times gone by. The Old Testament book of Jeremiah, for instance, recounts God's promise to make the descendants of David "as countless as the stars of the sky and as measureless as the sand on the seashore."

But in fact astronomers had, with some success, counted the stars. In the second century BC, the Greek astronomer Hipparchus set down the first systematic celestial catalog, listing the positions of about 850 of the brightest stars in the sky. Hipparchus divided the stars in his catalog into six groups according to their brightness. The brightest stars were assigned magnitude 1 and the fainter stars consecutively higher numbers, a system—with considerable modifications—that is still in use today. Hipparchus's catalog was updated about three centuries later by the Hellene philosopher and mathematician Claudius Ptolemy, who made his home in Alexandria, a cosmopolitan cultural center located at the mouth of the Nile. Ptolemy's great book on astronomy, the *Almagest* (which literally means "The Great Work"), was the most influential source of information on the heavens for the next 1,500 years. Galileo would have been very familiar with its handy list of 1,028 stars, which included virtually every object easily visible with the naked eye from his latitude. Some of the objects in Ptolemy's catalog were to prove especially interesting through Galileo's new telescope, especially those Ptolemy labeled as "nebulous," or fuzzy in appearance.

That there were more stars than those in Ptolemy's catalog was obvious from the start. About a third of the sky, including the stars far to the south, could not be seen from the lands bordering the Mediterranean. Aristotle himself cited reports of travelers who noticed new stars appear above the horizon as they traveled toward the equator. It was also clear that some observers were more sharp-eyed than others, and so the faintest stars could be seen by some people more easily than by others. Saint Augustine, admittedly no great expert on astronomy, commented in the fifth century AD that, when counting stars, "it is not to be believed that all of them can be seen. For the more keenly one observes them, the more does he see. So that it is to be supposed some remain concealed from the keenest observers."[14]

---

[14] Grant, Edward. *Planets Stars, and Orbs.* New York: Cambridge University Press, 1994. 444.

However many stars there were—and the number was surely at least a few thousand—most of Galileo's contemporaries regarded them as objects not unlike the planets. Aristotle, who was the ultimate authority on scientific matters, had not written clearly on the subject, but the prevailing view was that stars were large spheres made of some unique and strange celestial substance. Like the planets, they went around the Earth, and though there was some disagreement about whether each star was carried by its own individual sphere or whether they were all fixed like pushpins to a single sphere, the general view was that the stars were the most distant of the celestial objects, located in the uppermost region of the heavens. They moved as if part of a single unit, bound together in the fixed patterns of the constellations and linked to the outermost moving sphere of the universe which rotated west to east around its axis every 24 hours.

Astronomers in Galileo's time, with very few exceptions, believed that stars had no light of their own. Stars shone because they were illuminated by the Sun, just as the planets do. Starlight, however, was not simply reflected sunlight. According to the commonly held view, the stars were translucent or transparent objects, resembling giant snow globes cruising through the heavens. When light from the Sun fell on them, they filled with light, like glow-in-the-dark rubber balls, and reemitted the light in all directions. Such behavior may seem strange, but then according to Aristotle, the heavens were made of a special celestial substance that had properties unlike anything familiar in our terrestrial experience.

Of course, to catch enough sunlight to glow, the stars had to be relatively close to the Sun. Accordingly, whether or not they had doubts that the Earth was the center of the Universe, seventeenth-century astronomers were in agreement that the distance to the stars was of the order of tens of millions of miles, much closer than we know they are today. Tycho Brahe, for instance, the Danish astronomer who was the greatest observer of the heavens in the pre-telescopic era (he died in 1601), estimated the distance to the stars as about 14,000 times the radius of the Earth. Since the size of the Earth was fairly accurately known at the time, Brahe's stars work out to be a bit less than 60 million miles away—closer than the modern-day distance between the Earth and the Sun. It wasn't unreasonable, therefore, for seventeenth-century astronomers to believe the stars derived all their light from the Sun.

Astronomers of Galileo's era also reasoned that stars came in various sizes. Clearly this is so, by their logic, because not all stars are the same brightness. Those that were the brightest were supposed to be the biggest, while those that could barely be seen on a dark night were supposed to be smallest. Typical sizes for first-magnitude stars, like Sirius and Vega, were supposed to be a little over 100 times the diameter of the Earth, or about a million miles, and the faintest stars were supposedly about a dozen times the size of the Earth.

There is, however, one light in the nighttime sky that does not lend itself to such open-and-shut explanation, and that is the Milky Way. Four centuries ago, it was a more familiar sight to people than it is from today's light-polluted cities, but still, if you have ever gone camping in a relatively remote area, you know how impressive a sight it can be. It appears as a ragged, whitish cloud of light stretching from horizon to horizon. If you watch it closely, you become aware that it rises and sets with the stars. Though you can only see a portion of it at one time, after awhile you may come to realize that the Milky Way is not just a patchy splotch, but a band that encircles the entire sky. In the summer months, when it is high in the sky in the evening, its vast extent is particularly spectacular.

The Milky Way presented a challenge to astronomers of the past. Individual stars viewed with the naked eye seemed to group themselves into patterns, or constellations, that were identified with stories from ancient myth (Ptolemy recognized 48 of these constellations in his catalog). But when it came to the Milky Way, it was hard to see any such distinguishing pattern. Its edges seemed fuzzy, and it was uneven in brightness, with some very dark areas at its edges, like bays in an ocean, and little cottony patches of cloudy brightness here and there. Some of those cottony patches were distinct enough that Ptolemy considered them a special type of "nebulous" star. Other than that, however, astronomers spoke of the Milky Way as a whole and did not create maps of it or speak of its various patches and projections as individual features.

To Galileo's predecessors, the Milky Way was a unique puzzle. Aristotle had not regarded it as a celestial object at all. Like a comet, Aristotle's Milky Way was located in the Earth's atmosphere, closer to Earth than the Moon. The heat of the Sun, according to the philosopher, evaporated dry gases from the Earth and lifted them upward, where they were ignited by fiery substances in the upper atmosphere.

There was a glaring problem with Aristotle's explanation; it didn't seem to describe the Milky Way everybody knew. To anyone who looked at the sky very frequently, the Milky Way *did* look like a permanent feature of the celestial regions, rotating around the sky once every 24 hours just as the stars did. It wasn't a changeable meteorological phenomenon. Though they may have paid lip service to the great philosopher on most matters, knowledgeable astronomers in later centuries often adopted alternate explanations for the Milky Way, some of which had been knocking around since before Aristotle. The Greek Theophrastus, for instance, regarded the Milky Way as the seam where the two halves of the heavens were joined together; Diodorus of Cronos preferred to think of it as a belt of compressed fire—a notion that didn't seem to improve much on Aristotle's. And among the wild ideas, there was one that was fairly close to the mark—that of the philosopher Democritus, who is famous for being the first to suggest that all matter is made up of tiny atoms. Democritus explained the Milky Way as the light of a vast number of very faint stars, whose combined light gave the impression of a continuous strip across the sky.

Democritus pretty much had it right, and many astronomers in Galileo's time tacitly accepted his explanation of the Milky Way as the most plausible. But good science needs good evidence, and until Galileo's telescopic observations, there was no way to set the matter to rest. The true nature of the Milky Way, in the days before the telescope, was a question no more answerable than the question of how many angels could dance on the head of a pin.

Thus it was that on a clear, dark night in early 1610, Galileo turned his telescope on the Milky Way and on several of Ptolemy's "nebulous" stars. There was no question of what he saw: the silent glow of nebulosity, wherever it appeared to the naked eye, easily resolved itself into a multitude of stars through the telescope. It was so striking and so convincing that, even though the report on his observations of Jupiter and the Moon was already in press, Galileo quickly composed four additional pages of text on the stars and the Milky Way, which were bound, unnumbered, into the first editions of *Sidereus Nuncius*.

"The Milky Way itself," he wrote, "with the aid of the spyglass, may be observed so well that all the disputes that for so many generations have vexed philosophers are destroyed by visible certainty, and we are liberated from wordy arguments. For the Galaxy is nothing else than a congeries of innumerable stars distributed in clusters." The telescope easily showed a giant cloud of individual faint stars stretching across the sky. Telescopic evidence was proving the truth of the old Biblical prophecy: the stars in the heaven were indeed countless. Even a few objects that Ptolemy had classified as single "nebulous" stars in his catalog turned out to be "swarms of small stars placed exceedingly close together." Galileo noted that Praesepe (commonly called "the Beehive"), a fuzzy object in the constellation Cancer, turned out to be not one, but a cluster of more than 40 faint stars.

To Galileo's mind, and to those of many of his contemporaries, the telescope had settled the matter. The Milky Way was a celestial object containing a vast number of stars. There were also myriads of previously undetected individual stars that could be seen through the telescope in other parts of the sky. Galileo noted in particular his observations of the Pleiades, or the "Seven Sisters," a tight star cluster that many novice skywatchers mistake for the Little Dipper. (It appears on the hood insignia of Subaru vehicles; "Subaru" is the Japanese term for the Pleiades.) Where naked-eye observers had charted six or seven stars, Galileo could spot nearly 50.

Still, the telescope left unanswered all ambiguities and disputes about the true nature of stars. Galileo was surprised to find that the stars didn't look much different through the telescope than they did to the naked eye. They were brighter, of course, but unlike the planets, which looked notably disklike when magnified, the stars still looked like tiny points of light. In fact, the brightest stars, like Sirius, which astronomers considered to be the biggest, presented themselves as no larger through the telescope than did the faintest stars, which were supposed to be at least 10 times smaller.

The reason that the stars were not effectively magnified, we know today, is that they were too far away to be resolved by Galileo's telescope. The farther away an object is, the smaller an angle it appears to cover in the sky (this called the *apparent angular size* of an object). When a telescope magnifies, it makes that angle appear bigger. But there is a minimum angular size that a terrestrial telescope can magnify, called the *resolution limit* of the telescope. If an object has an apparent angular size below the limit of a telescope, no amount of magnification will make it appear as anything other than a featureless point of light. Planets have large enough angular sizes to be resolved, even though they are actually much smaller in diameter than most stars, and that is why Galileo was able to see them as circular disks. But even the closest star beyond the Sun has such a small angular size that it appears as just a featureless point.

What were these stars? How far away were they, really, and what made them shine? In his first rushed-to-press report in *Sidereus Nuncius*, Galileo didn't speculate. He just reported on the strange and remarkable things he saw through the telescope. Even in the matter of the Milky Way, though he was quick to claim to have proved it was made up of faint individual stars, he didn't ask how far away they were or why they seem to be confined only to a narrow strip in the sky.

Such speculation, he may have realized, could be dangerous. Ten years earlier, Giordano Bruno, a maverick priest with a penchant for cosmological invention, had been executed for heresy. Among his many unconventional views (though not the principal cause of his execution), was the notion that the stars were suns like our own Sun. Each star, he proposed, had its own retinue of planets and probably its own inhabitants. Breaking with long tradition, Bruno proposed that the stars were not confined to a single sphere but were scattered randomly throughout an infinite universe.

Copernicus was a major inspiration for these views, but Bruno was no scientist. He supported his ideas with philosophical arguments and appeals to the mystical qualities of nature, not with observational evidence. There was no evidence to cite, in fact, for Bruno wrote his two major works in 1584, a quarter of a century before Galileo constructed his first telescope. Nevertheless, Galileo must have been aware that, even with the "hard data" of telescopic observations to back up his own claims, talking about distant solar systems and self-luminous stars might still be very risky business.

Later in his life, after his telescopic observations had pretty much ended, Galileo wrote his eloquent and controversial defense of the Copernican system, *Dialogue on the Great World Systems*. While the book deals primarily with the question of whether the planets go around the Earth or the Sun, there are a few passages that reveal Galileo's thinking about the stars. The stars, he seemed to believe, were located at huge distances from the Sun, much farther than the tens of millions of miles that astronomers had previously believed. That explained why they could not be resolved with his telescopes. Different stars were at different distances from us, though the Universe was still bound by a huge sphere and was not infinite in extent, as Bruno—God forbid—had suggested. And it seemed likely that stars shone of their own light: after all, they were far too remote from the Sun to be illuminated directly by it. Independently of Galileo, other imaginative astronomers like Johannes Kepler had reached similar conclusions.

All this, however, was blue-sky invention, given the scanty evidence provided by the earliest telescopic observations. Centuries would pass before astronomers succeeded in directly measuring the distance to stars, and even more centuries would pass before they were able to detect the presence of planets around other stars. Galileo's pioneering telescope could dispel a few ancient notions about the Milky Way in the early 1600s, but when it came to revealing the true nature of stars, it could only point the way onward and upward.

THE STARS AND THE MILKY WAY TODAY

Galileo found that the visual phenomenon of the Milky Way—the broad band of dim white light across the night sky—is a result of the cumulative presence of innumerable stars, most too faint to be seen individually with the naked eye. Today we identify this visual Milky Way as our view of a vast, relatively flat system of hundreds of billions of stars, our Milky Way galaxy, from our location within it. And we recognize stars as naturally occurring nuclear furnaces, burning the fuel in their interiors.

Just as a campfire burns lower as wood is consumed, we know that stars burning their internal fuel progress through stages in which they may change in size, temperature, and other physical properties, forging heavier elements from lighter ones and often ejecting much of their constituent matter to the interstellar medium, the gas and dust present between stars in the Galaxy.

Stars are born from gas in the Galaxy and progress from one stage to another, forced to change by the nuclear reactions inside them. The Galaxy, which formed from a vast cloud of primordial gas, also evolves. It develops as stars are born, return some of their gas to the interstellar medium, and die, and also by absorbing smaller galaxies that formed near the Milky Way or come too close to it.

The stars seen through modern telescopes come in a great variety. There are red giant stars and white dwarf stars, red dwarfs and blue supergiants, and more. There are variable stars that pulsate in and out rhythmically, changing in size, brightness, and temperature over and over again. Other stars have huge stellar flares, magnetically powered explosions that increase their brightness by a noticeable amount for a short time, and then they subside until the next flare.

We identify each of the different types of stars with a particular stage in stellar life. Protostars, hidden from view within dark clouds in the Milky Way but detected with infrared and radio telescopes, turn into stars like the Sun, which burn hydrogen in and near their centers. Stars that resemble the Sun in this way but are more or less massive have different colors. The Sun has an off-white tint, while stars twice as massive are pure white, stars a little less massive are yellow or orange, and stars much less massive are red and small. The colors of the stars are directly related to their surface temperatures. Stars more massive than the Sun are white or, if sufficiently massive, blue-white, and are much hotter than the Sun. Stars that are yellower than the Sun, or orange or red, are progressively less hot.

Stars generate their own energy until they run out of fuel, and then, in most cases, they slowly fade away. All the stars that we can see with the naked eye are generating their own energy by nuclear fusion within them. The stars that have run out of fuel are too dim to be seen without a telescope.

There are even objects like stars, but dimmer, cooler, and less massive, that do not conduct nuclear reactions, or do so at most for a short time. They are the so-called brown dwarfs. A brown dwarf is about the same size as the planet Jupiter but much heavier—from about 13 times to about 80 times more massive than Jupiter.

Some stars, having exhausted their nuclear fuel, are barely shining if at all. They have shrunk down to very small sizes and have become extremely dense as a result. Although small, their gravity is extremely strong. These objects include white dwarf stars, which are roughly as big as the Earth but far more massive than any planet, with typically about 60 percent as much mass as the Sun. White dwarfs are so dense that a teaspoon of their material would weigh a ton on Earth. A white dwarf is shining because it is hot. However, as it is not producing new energy, it slowly fades away over millions of years, like a very long-lived ember glowing in the remains of a campfire.

Despite its small size and great density, a white dwarf is bloated and airy in comparison to another kind of star whose nuclear burning has ended: the neutron star. A typical neutron star is no larger in diameter than a large city on Earth but is 40 percent more massive than the Sun. That huge mass is packed so closely together that a teaspoonful would weigh 1 billion tons on Earth. It's so small that it produces very little light that we can see.

A neutron star is so hot that it shines predominantly in X-rays, which gradually fade as it cools. However, some neutron stars are highly magnetized and spinning rapidly, making a complete turn in less than one second or, in extreme cases, turning hundreds of times per second. They produce beams of radio waves (and in rare cases, visible light waves). As the neutron star turns, a beam may sweep across the direction of the Earth, and we receive a short pulse of radio waves as the beam rotates by. These beamed stars are called pulsars. The most famous neutron star is the pulsar in the Crab Nebula. It is a rare case in which a pulsar also has beams of visible light, so we can see it flashing on and off in time-series telescopic photographs of the Crab. (The Crab Nebula is the gaseous remains of a supergiant star that exploded in the year AD 1054 and was seen in China, Japan, and elsewhere. The stellar explosion, or supernova, was so bright for a short period that it could be seen in broad daylight.)

There's no equivalent on Earth to the dense matter in a white dwarf or the denser stuff in a neutron star, but a third kind of "end state of stars"—what stars become when their nuclear fuel has run out—is even weirder. It is the black hole.

A black hole is formed when a star, no longer supported by the pressure of heat and radiation generated by nuclear reactions in the interior, collapses under the force of its own gravity, with nothing to ever stop it. White dwarfs and neutron stars are formed by collapse as well, but in each case, a state of dense matter is reached that develops enough pressure to resist further collapse. A collapsing star that will produce a black hole is so massive, however, that there is no force at all that can stop it from falling inward. There is a remarkable consequence: the *escape velocity* of the object becomes greater than the speed of light. The escape velocity is the speed needed to depart a celestial body forever. For example, a rocket bound for Mars must take off with a velocity of at least 7 miles per second, the escape velocity of Earth, to get away from our planet. But once the escape velocity of an object exceeds the velocity of light, which travels at 186,000 miles per second through the vacuum of space, then no rocket can get away. In fact, nothing can get away from the black hole—not even a ray of light.

Matter can fall into a black hole, but it can't get out. It is then defined as falling within its *event horizon*, the distance from the center at which the escape velocity equals the speed of light (outside the horizon, the escape velocity is slower than light speed). For a black hole with 10 times the mass of the Sun, the event horizon is 37 miles in diameter.

Stars that form with masses of less than eight times the mass of the Sun (and that are not at the low end of the mass range, like red dwarfs) will eventually become white dwarfs or neutron stars; the ones with lower masses become white dwarfs, and the heavier ones turn to neutron stars. Stars with more than eight times the mass of the Sun become supernovae. Among them, the smaller ones produce neutron stars, and the more massive ones become black holes.

Stars also come in pairs and groups. About half of all stars in the Milky Way have one or more companion stars. There are binary stars, consisting of two stars that were born together (in most cases) and that each orbit a point in space between them (their "center of mass"). Triple stars consist of a binary star and a third star in a much larger orbit, all moving around their common center of mass. There are also quadruple stars, usually consisting of two binary stars; each binary system follows an orbit around the common center of mass of the whole quadruple system. Other multiple-star systems can have five, six, or more members. Even larger groups are known as star clusters.

We can't see a black hole directly, since light cannot get away from it, but we can detect it through its effects on nearby matter, such as a companion star. That's how black holes sometimes are discovered and studied—a star is found moving back and forth as though it is in orbit with a binary companion of considerable mass, but no massive star is observed.

The Pleiades that Galileo studied is a so-called *open star cluster*. We can easily see it with the naked eye in fall or winter in the Northern Hemisphere in the constellation Taurus. When we look at it through binoculars or a telescope, we see many more stars than with the naked eye, just as Galileo did. Open clusters have dozens, hundreds, or even thousands of stars. Some of them are literally flying apart, with insufficient gravity to hold them together. These open clusters, seen in the act of breaking up, are also called stellar associations. Stellar associations are only seen when they are a few million years old or so, because not long after that, they come apart.

There's another type of star cluster that is found in abundance in our galaxy. They are massive ball-shaped collections of tens of thousands, hundreds of thousands, or even millions of stars. When we look at one of these *globular star clusters* through a telescope, we see the big difference between them and open clusters (other than that a globular cluster contains many more stars): the stars in an open cluster are loosely spread out over a region of space, while in a globular cluster, the stars are packed more and more closely together toward the center.

To astronomers, the important thing about star clusters of all types is that all the stars within a cluster formed from the same initial cloud of material at essentially the same time. So by studying a cluster, we can learn the average properties of different kinds of stars of the same age. And by studying a variety of star clusters, we learn the properties of the same kinds of stars at different ages. It's statistical research like this that enables us to deduce the life cycles of stars.

THE LIVES OF THE STARS

One common thread connects all the different types of stars: by and large, the nature of a star and the stages it goes through during its life depend on its initial mass, or the amount of mass with which it formed. The amount of energy that a star emits, its color and temperature, the stages it goes through (that is, if it resembles the Sun, or is a red dwarf, or becomes a red supergiant, and so on), and even the star's life expectancy are determined by its initial mass.

During most of its nuclear burning life, a star is fusing hydrogen into helium in its deep interior, where the temperature and density are so great that four hydrogen nuclei (each is simply a proton) combine to form the nucleus of the next-heavier element, helium. The helium nucleus created in this way is not exactly four times as massive as four hydrogen nuclei, however. It is a tiny bit lighter, and this difference of about seven-tenths of a percent represents mass that has been transformed to energy by the nuclear reactions. That energy liberated in the fusion process by the destruction of a small amount of matter is what powers the star.

A big blue-white star with 60 times the mass of the Sun shines furiously, pouring out almost 1 million times as much energy as the Sun every second, and is about 20 times as large in diameter. But that stellar powerhouse runs through its nuclear fuel in less than a half-million years. The Sun has been shining about the same as it does now for 5 billion years and will continue to do so for about that many more years before it goes through a major change.

On the other hand, a tiny red dwarf with a few percent of the mass of the Sun and a small fraction of its size may shine at less than a tenth of 1 percent of the rate at which the Sun does, but it will do so for hundreds of billions of years to come, long after the Sun has disappeared from view. Proxima Centauri, the nearest star beyond the Sun, is a red dwarf. Even though it is closer to Earth than any star that you can see with the naked eye, Proxima is much too dim to be seen without a telescope.

The Sun is more massive than the great majority of stars, so most stars will live longer than the Sun, but they shine less brilliantly and at cooler temperatures. However, there are exceptions to the general rule in which the mass determines the state and the future of a star. When two stars are born with the same mass but with different chemical compositions, they will develop somewhat differently. In rare cases in which two stars in a binary system are so close to each other that mass can transfer from one star to the other, the nature of the stars is affected. In certain cases, matter falling down onto one star in a binary system from the other can trigger nuclear fusion of hydrogen at the surface of the mass-receiving star, and the fusion will sometimes produce a runaway nuclear reaction that blows up the whole star in another type of supernova. This explosion doesn't leave a neutron star or black hole behind; it shatters the entire star and throws all of its material out into space at high speed.

The Milky Way contains thousands of clouds of gas and dust called nebulas. The Orion Nebula is the most famous one; it is easily seen with the naked eye in the fall and winter evening skies in the

Northern Hemisphere and in the spring and summer in the Southern Hemisphere. Galileo must have seen it since he looked at stars in the constellation Orion with his telescope, but there is no record that he examined the nebula.

The Orion Nebula is a bright cloud that contains a cluster of hundreds of young stars that were born at nearly the same time. They were born in a much larger nebula, the Orion Molecular Cloud, which is, for the most part, so cold and dark that it is not visible to the eye. As they formed and began shining brightly, they began to heat the nebular gas in their vicinity and make it expand, thin, and glow. The small region of the Orion Molecular Cloud that is shining is the Orion Nebula, a hot bubble on the edge of a great dark ocean of cold gas.

The stars formed in the Orion Molecular Cloud as they do elsewhere in the Milky Way, as the densest, darkest, and coldest clumps of gas succumbed to the inward-directed force of their own gravity. Observations made in infrared light and radio waves, which can penetrate the dust that obscures molecular clouds from ordinary telescopic view, show that groups of clumps shrink until each is a few trillion miles across. These denser clumps, called "cloud cores" by astronomers, are past the point of no return: each must fall in on itself, the infalling gas dropping faster and faster, until something begins to stop their collapse. The force or effect that stops the collapse of a forming star, or protostar, is pressure. The more you compress a gas, the higher the pressure becomes and the hotter it gets. Gas pressure, as well as the pressure of radiation—the light generated in the hot gas—act to expand the gas, opposing the inward pull of gravity. In other words, when the center of the protostar gets so hot and dense that nuclear reactions begin, it generates enough energy to increase the pressure to the point at which it stops the collapse. Now the protostar has become a normal star, which burns hydrogen to helium, like the Sun. It takes about 10 million years for a star with the mass of the Sun to go through the protostar phase but only about 100,000 years for a star with 10 times the mass of the Sun to collapse and start nuclear burning.

The birth of a star through gravitational infall in a cloud of interstellar gas does more than create the star. It often produces planets as well. Astronomers have discovered over 300 planets of stars in the Milky Way and beyond the Sun, called exoplanets. The great majority have about the same mass as Jupiter or more mass, because heavier planets are easier to find with the most common telescopic detection methods. However, lighter planets have been found, down to the mass of Saturn and below. As we discussed in Chapter 6, we believe that it is only a matter of time before technology allows us to discover rocky, possibly water-bearing planets with the same mass as the Earth.

Many of the exoplanets that we know about so far are in detectable solar systems or planetary systems of their own. In other words, they are part of a system of two or more planets, all circling the same star. As of early 2008, an exoplanetary system with at least five member planets has been found and it is just a matter of continuing the observations until additional members are found in this or other systems.

THE MILKY WAY GALAXY

The Milky Way is a spiral galaxy, with a flattened disk of stars, gas, and dust, including thousands of bright and dark nebular clouds, open star clusters like the Pleiades, and a central bulge of mostly older stars. It also has a huge spherical region, the galactic halo, that reaches great distances above and below the disk and contains hundreds of globular star clusters.

In the disk, the clouds of gas are distributed in a pattern of about a half-dozen prominent spiral arms, a bit like the streams of water from a rotating lawn sprinkler. The more massive young stars, blue and white, and the open star clusters all form in the spiral arms and outline them like bright lamps among the gas and dust.

The disk is about 100,000 light-years across and about 1,000 light-years thick. The galactic halo extends over a diameter of about 300,000 light-years. The Sun and its planets, including Earth, are located in or near the Orion spiral arm at about 25,000 light-years from the central point in the Milky Way, the galactic center. Everything is orbiting around the galactic center, and at the Sun's distance, it takes 226 million years to make one complete circle, a period of time called the galactic year. However, the orbiting is not confined to a plane. Attracted by the gravity of large numbers of stars above and below the central plane of the disk (the galactic plane), the Sun and planets move up and down as they orbit around the center, like riders on the bobbing horses of a carousel. Each of these up-and-down cycles probably takes about 60 million years, so there are three or four bobbing cycles in every galactic year.

In the innermost region of the galaxy, a large number of stars seem to rotate around the galactic center in an elongated swarm that, viewed from the right angle, is like the spoke of a wheel. That spoke is the galactic bar, about 15,000 light-years long from tip to tip.

In the central few light-years of the Milky Way, stars are moving at remarkably high velocities; one star is swinging around the galactic center at over 3 million miles per hour. They are moving under the gravitational sway of a so-called supermassive black hole. It's not the remains of an exploded star like the black holes we have discussed, because its mass is far greater than that of any possible star; it is almost 4 million times the mass of the Sun. It must have formed early in the history of the Milky Way, perhaps as part of the process in which the Galaxy as a whole collapsed. Or it may have preceded the existence of the Milky Way and perhaps was somehow involved in inducing the formation of the Galaxy.

The Milky Way contains hundreds of nebulae like the Orion Nebula, which glow because of energy from young stars that formed within them. There are also many reflection nebulae, gas clouds that shine because the dust particles within them reflect the light of nearby bright stars. And there are many so-called planetary nebulae, which are puffs of gas ejected from stars like the Sun late in their lives. By that point, there's nothing left of the star (as there will be nothing left of the Sun) but a very small, very hot object, the remaining core of the red giant that the star became. The vast outer regions, or atmosphere, of the red giant have blown away in stellar winds over millions of years, and some large puffs have ejected gas that is heated and made to glow by ultraviolet rays from the small, hot star. The Ring Nebula is one of the many beautiful examples of planetary nebulae, which have nothing to do with planets but sometimes look like a dim, greenish planet, like Uranus, in a small telescope at low magnification. After a while, the planetary nebula produced by a star is gone, dissipated into space, and the little star at its center has cooled off and will keep cooling forever, shedding its residual heat. It has become a white dwarf.

There are also many nebulae in the Milky Way that were produced by supernova explosions, like the Crab Nebula. These "supernova remnants" return much of the mass of an old star to the interstellar medium of the Milky Way. Stellar winds, especially from red giant stars, return mass as well.

Because new, heavier elements are forged inside stars by nuclear reactions, the gas that is returned to interstellar space by stars has a different chemical composition than the gas from which the stars first formed. It has larger amounts of elements heavier than hydrogen, like carbon, oxygen, and iron. So as time goes on in the Milky Way, new stars form from the interstellar gas clouds, which have been enriched with heavy elements, and thus the chemistry of the Galaxy changes over time.

Another process that makes significant changes in the nature of the Milky Way and other galaxies is mergers with other galaxies. The Milky Way has several much smaller galaxies that are in orbit with it, called satellite galaxies, and in some cases it is pulling them apart by tidal force. There are also several known star streams in the Milky Way that seem to consist of stars that once made up part of another small galaxy that may have been completely disrupted over time. Their identity as members of a stream that originated in this way is shown by their motions and chemical composition, which have much in common with each other but are different from the surrounding stars of the Milky Way.

The Milky Way is more than 12 billion years old, but it has changed greatly over time, and it is still changing.

CHAPTER 10

Cosmology

COSMOLOGY IN THE SEVENTEENTH CENTURY

Galileo's first telescope did far more than add a few amusing facts to our knowledge of the Universe; it changed the course of science. Granted, your average seventeenth-century reader of Galileo's *Sidereus Nuncius* and *Letters on Sunspots* must have enjoyed Galileo's descriptions of celestial wonders: the multitude of faint stars in the Milky Way, four moons circling like clockwork around Jupiter, the strange shape of Saturn, and the rotation of dark spots on the Sun. But at the same time, it was impossible for any literate European to ignore the revolutionary implications of Galileo's discoveries. They challenged the most accepted truths of science and even threatened the authority of the Church.

Revolution was in the air. The Protestant Reformation, already under way for almost a century, had cast cherished beliefs and institutions into doubt. New ideas and new critiques of the establishment were sparking furious debates in intellectual circles. Science was awash in a rising tide of skepticism and speculation.

In the century preceding Galileo's first observations, a number of independent thinkers had called into question long-held notions about the nature of the cosmos and the basic physics of Earth and sky. Some of the issues were relatively clear-cut: Was the Earth or the Sun the center of the Universe? What caused the planets to move around the sky? Others issues, more philosophical or theological in nature, sound strange to modern ears: Could the workings of nature be expressed in mathematical language? Could purely physical laws effectively describe the order of nature? Did planets and stars have animate souls whose motivations and desires governed how the natural world behaved?

As we have discussed in previous chapters, one of the more controversial scientific players in the debates of the sixteenth century was Nicolaus Copernicus. In his book *De Revolutionibus Orbium Coelestium*, he proposed an alternative to the prevailing view that the Moon, Sun, and five planets circled the Earth. According to Copernicus, only the Moon went directly around the Earth, while the Earth and the other five planets orbited the Sun, which was located at the center of the Universe.

Copernicus's *heliocentric* (Sun-centered) system was a radical break from the *geocentric* system, the Earth-centered picture of the heavens devised in the fourth century BC by Aristotle. Aristotle's geocentric Universe consisted of a series of nested crystalline spheres, a sort of cosmic Russian *matryoshka* doll, whose revolutions produced the observed motions of the heavens. The innermost sphere carried the Moon, followed by spheres for Mercury and Venus, the Sun, Mars, Jupiter, Saturn, and finally the fixed stars. The outermost star sphere rotated once every 24 hours, bearing with it all the spheres inside and thus producing the daily rising and setting of the stars, Moon, Sun, and planets. To account for some of the more complex motions of the heavens, like the occasional reversal of direction of the planets, Aristotle added a system of secondary spheres between the main spheres, tilted at various angles. In all, his system required 55 spheres for the stars and 7 planets.

Aristotle's concentric spheres became the standard image of the Universe, but they were not very useful for practical matters, like predicting the exact positions of the planets or the dates of lunar eclipses. In the second century AD, Claudius Ptolemy devised a more mathematical version of the system, retaining Aristotle's notion that motions in the heavens were circular and Earth-centered but dispensing with the many cumbersome spheres. In the Ptolemaic system, the planets traveled around the centers of imaginary circles called epicycles, while the centers of the epicycles circled the Earth in paths called deferents. The calculations, however, especially in the days before calculators, could be tedious, and medieval astronomers devised a number of clever shortcuts—such as rotating paper disks with angles marked on them—to make it easier to produce tables of planetary positions. "Had I been present at the creation," commented King Alfonso the Wise of Spain in the thirteenth century, when shown how astronomers calculated these tables, "I would have suggested something simpler."[15]

The Copernican system might have been the answer to King Alfonso's wishes, because it described the motions of the heavens more economically than Aristotle's orbs or Ptolemy's circles upon circles. The rising and setting of the Sun and stars, in Copernicus's Universe, was simply a result of the rotation of the Earth, not the motion of a huge crystalline sphere. And the occasional retrograde motion of a planet could be explained by the fact that the Earth was moving faster around the Sun than the planets farther out, so that the line of sight to a planet beyond the Earth appeared to reverse direction whenever the Earth passed between it and the Sun.

Yet astronomers were slow to accept the Copernican system. "The Earth feels so solid and motionless," was a common objection. "How could it spin around daily, much less travel yearly around the Sun?" The most serious criticism of the heliocentric theory, however, was not that it violated common sense but that it contradicted the fundamental laws of physics—in particular, Aristotle's laws of motion. According to Aristotle, the way a body moved depended on what it was made of. Aristotle visualized the Earth as being

[15] Koestler, Arthur. *The Sleepwalkers*. New York: Macmillan, 1959. 69.

made of one type of material, while the heavens were made of a different substance. This accounted for the difference between the motions of things close to Earth, which were mostly up and down, and things in the heavens like planets and stars, which were round and round.

In the terrestrial, or sublunar realm—the region inside the sphere of the Moon—there were only four basic elements: earth, air, fire, and water. Everything we see in the terrestrial world was a mixture of these four elements, and that accounted for why heavy bodies like rocks fell to Earth, while light bodies like smoke and flame floated upward. Aristotle's explanation was couched in terms that sound more organic than physical: the elements have natural resting places, and if they are separated from these resting places, they head back home at the first opportunity. So a rock falls downward because it contains a lot of earthy material, whose natural resting place is the center of the Universe (i.e., the center of the Earth). When dropped, it "wants" to get to its natural resting place, so it heads downward. Similarly, a flame leaps up because it is made mostly of fiery substance, whose natural resting place is the uppermost terrestrial region just below the sphere of the Moon. If Aristotle were teaching today, he might tell his classes to visualize the terrestrial world as a shaken-up bottle of salad dressing: if left to settle out, it would separate into its constituent layers—earth at the bottom, then water, fire, and air.

In contrast to this terrestrial up-and-down tendency, objects in the heavens—the Moon, Sun, and other planets—behaved differently from objects near the Earth, as anyone could see. Celestial objects moved *around* the center of the Universe, neither falling downward nor floating upward. That, said Aristotle, is because they contain neither earth, nor air, nor fire, nor water. Everything from the Moon on out to the stars is composed of a fifth element, the quintessence or aether, whose natural tendency is to move in a perfect circle. Given sufficient time, a heavenly body always returns to exactly the same place from which it started, tracing out the same circular path over and over again. Thus, everything in the heavens can be considered perfect, changeless, and eternal.

Aristotle's explanation of motion was extremely convincing, dominating science in Europe for almost 2,000 years. It's easy to understand why. Change is all around us: things are born and die, trees fall, and rivers flow. Amid all the various motions and alterations in everyday life, Aristotle's laws of physics revealed a kernel of stability. The motions of terrestrial objects may have looked random and chaotic to the untrained eye, but Aristotelian natural philosophers could see fundamental laws operating in every natural event. Furthermore, as a backdrop to the infinite changeability of the terrestrial realm, there was the perfect regularity of the heavens, the eternal cycling of stars and planets that revealed a divine order to the Universe.

Though Aristotle set down his ideas three centuries before the birth of Christ, Christian scholars who rediscovered his work in the Middle Ages adopted the divine order of the celestial spheres as a model of the kingdom of God. The changeable and physically "corruptible" terrestrial realm below the Moon was considered the appropriate place for morally corruptible humans, who carried with them the sin of Adam. The system of nested shells mirrored the hierarchy of the Catholic Church, and the perfection of the heavens reflected the infinite power and goodness of God. In fact, in the many medieval depictions of the Universe that appeared in books and paintings, the region beyond the sphere of the stars is the Empyrean Heaven, populated by angels and by God Himself, whose power provides the motive force to keep all the cosmic spheres in motion.

So though Copernicus's Sun-centered Universe may have reduced some of the complexities of the old Aristotelian/Ptolemaic geometry, it raised an even greater number of troublesome questions about physics and theology. If the Sun were the center of everything, why then did rocks fall down, and why did smoke rise? Why didn't rocks fall toward the Sun if it was the center of everything? And by the same token, why does the Moon circle the Earth, but the planets circle the Sun? Why would God have created a Universe in which humans only inhabited a little world off to the side of the center of everything, when the Bible tells us that man was created in His image?

Though these knotty issues were implicit in the heliocentric system, Copernicus's ideas did not raise much of an uproar in the years immediately after *De Revolutionibus*. The Copernican Universe was so out-of-step with the current understanding of nature and so unsupported by any compelling evidence that it hardly posed a viable threat to established teachings. Had there not been other, more substantial reasons to doubt Aristotle, we might still be debating whether he or Copernicus had the better grasp on truth.

But by the time of Galileo, there was growing disenchantment with orthodox cosmology, largely because of new discoveries made late in the sixteenth century. In November of 1572, a new star, or nova, appeared in the constellation of Cassiopeia. Such new stars were seen every century or so, frequently enough to be objects of scientific speculation but sufficiently rare that there was practically nothing of substance known about them. Aristotelian philosophers interpreted these new stars as atmospheric phenomena, a sort of condensed fire like a bolt of lightning, but they based their conclusion more on logic than on evidence. A new star could not be farther than the Moon, they reasoned, because the heavens never changed, at least according to Aristotle's laws of physics.

A young Tycho Brahe was prepared to put the new star to the test. In Brahe's time, several decades before the introduction of the telescope, astronomers could only measure the angles between things in the sky. Even with these limited capabilities, however, it was possible to tell whether an object was closer than the Moon or farther away. The key was parallax, caused by the change in perspective of an observer as an object in the heavens rotated around him during the night. That change in perspective would cause an object closer than the stars to appear to shift in position against the star background. One could even tell how far away the object was, because the closer the object, the larger the shift. (You can see this shift yourself by holding your thumb at arm's length, closing one eye, and lining the thumb up with a distance object. If you switch eyes, the thumb jumps, and if you bring the thumb closer, it jumps more.)

Brahe used a new sextant he had just purchased to measure the separation between the nova and some of the other stars in

Cassiopeia. He expected to see a shift of more than two degrees over the course of a night, if indeed the star were closer than the Moon. But the star refused to budge—not on Brahe's first night, not a month later, and indeed not anytime during the 18 months that it remained visible, finally fading from view. Brahe had to conclude that the star was far beyond the Moon, possibly as far away as the outermost sphere of stars. A few years later, in 1588, he found a similar lack of parallax for a bright comet. The evidence seemed to be proving Aristotle wrong on at least one important point; there were things in the heavens that changed, just like things on Earth.

Brahe's observation of the nova of 1572 marked the beginning of his career as a great observer of the heavens. He became so well known for his writings on astronomy that, in 1576, King Frederick of Denmark granted him an island, Hveen, in the strait that nowadays separates Denmark and Sweden. There Brahe established an observatory, equipped with exquisite instruments of his own design, and for the next quarter of a century, he kept a continuous record of the positions of the planets in the sky. The nova of 1572 had shaken his faith in Aristotle, but he realized that the only way to come up with a better system was to compile hard evidence about how the planets really moved. "I am certain that there are no solid spheres in heaven," he wrote in 1590, "no matter if these are believed to make the stars revolve or to be carried around by them."[16] His life's work was to prove Aristotle wrong.

Alas, life is short, and work is long. Faced with advancing age and the difficult task of interpreting the mountain of numbers he had collected, Brahe hired a young mathematician named Johannes Kepler to ensure his legacy. Kepler's task, as Brahe saw it, was to demonstrate mathematically that the observed motions of the planets favored Brahe's pet model of the Universe, which featured a stationary Earth, an Earth-orbiting Sun, and a retinue of planets orbiting the Sun. When Brahe died in 1601, however, Kepler found that neither the Tychonic system nor the old Ptolemaic system could reproduce the planetary

---

[16] Marschall, Laurence A. *The Supernova Story*. Princeton: Princeton University Press, 1994. 81.

motions that Brahe had so carefully recorded. The only explanation for Brahe's observations was to put the Sun at the center of the planetary system, as Copernicus had suggested six decades earlier.

Still Kepler could not match Brahe's observations with a "pure" Copernican system, which shared with the Aristotelian system the notion that planets move in perfect circles at uniform speed. The only orbits that seemed consistent with the numbers from the real world required the planets to loop in close to the Sun, speeding up as if they were rolling downhill, and then swing away from the Sun while slowing down again. Geometrically, planetary orbits were elliptical, and the speed of a planet was inversely proportional to its distance from the Sun. Aristotle, it appeared, had been proven wrong again: things in the heaven could move in noncircular paths and at ever-changing speeds. Trustworthy old laws of motion seemed to be showing their age.

Kepler's book on planetary motion, *Astronomia Nova* (*New Astronomy*), came out in 1609, about the time that Galileo was making his first telescopic observations. News traveled slowly in those days, and Kepler, then living in Prague, first heard about Galileo's telescope from a friend in March 1610, just before *Sidereus Nuncius* was published. Kepler, an expansive thinker who always saw the big picture, was quick to understand that if Galileo had really seen satellites orbiting another planet and if there were really mountains on the Moon, Aristotle's Universe was in deep trouble. Kepler immediately tried to get his hands on a suitable telescope, and though there were none in Prague that were good enough, he managed to borrow one from a visiting German nobleman and make his own observations of Jupiter's satellites in August 1610.

Other astronomers, more wedded to the old Aristotelian system, were still reluctant to believe what they saw with their own eyes. Some granted that they saw new things through the telescope but doubted that these were anything but optical illusions produced not by real objects in the heavens but by reflections inside the telescope tube.

What old-guard Aristotelians found so hard to accept was that, if the telescope were to be trusted, it added a decisive element to the debate over whether Aristotle or Copernicus was correct. This was the most serious challenge that orthodox astronomy had ever faced, despite recent assaults by unconventional thinkers. Coperni-cus's system could be regarded as a mere "hypothesis"—clever, yet little more than a debating trick. Kepler's analysis of orbits could be ignored as highly mathematical and so dense that only a few obses-sive numerologists could understand it. Galileo's observations, on the other hand, were crystal clear, and he wrote as lucidly and com-pellingly as a skilled trial lawyer.

In fact, if he had been putting together a legal brief, indicting Aristotle for grievous errors in his description of the cosmos, Galileo might have brought forth the following points, based on his obser-vations between 1609 and 1613:

* The satellites of Jupiter orbit around an object other than the Earth, and they are not left behind as Jupiter moves. Clearly the Earth does not have to be the center of all heavenly motion. And just as clearly, the Earth could carry the Moon with it as it orbits the Sun.

* The Moon looks a lot like the Earth. It doesn't seem to be composed of an aetherial substance. Maybe the heavens are composed of the same stuff as the Earth.

* Venus shows a complete cycle of phases from thin crescent to full, just as predicted by Copernicus's system. In Aristot-le's and Ptolemy's systems, Venus should show only a limited range of phases and should never appear full.

* The stars are not magnified by the telescope. Maybe the Uni-verse is much larger than Aristotle imagined.

* The Sun is not "perfect" but shows spots that appear and disappear, like clouds on Earth. Moreover, the Sun rotates, making it equally plausible that planets, and the Earth itself, can rotate, despite the fact that they are huge objects.

Galileo certainly regarded these arguments as open-and-shut. Before 1609, the clean logic of Copernicanism may simply have appealed to his intellect, but what he saw through telescope touched him to the soul. He was now convinced that Copernicus had set forth the only true system of the Universe, and he expected others, once shown the evidence, to embrace heliocentrism as zealously as he did.

He was to be disappointed and frustrated. There were powerful advocates of the orthodox astronomy, especially among Jesuit astronomers in Rome, who would not give up without a fight and who had the ear of the Pope as well. For instance, after some initial questioning of the telescope itself, Christoph Clavius, one of the most influential of the Jesuits, granted that its revelations were not optical illusions. Yet Clavius continued to hold out hope that they could somehow be explained without scrapping Aristotle's ideas altogether, and he regarded Galileo as an impertinent dilettante.

The Jesuits were fighting for a losing cause, but the outcome of the scientific revolution was not to be decided in Galileo's lifetime, and the initial battles went badly for the new astronomy. In 1616, the Inquisition summoned Galileo to Rome to defend himself against charges that he was teaching that the Earth moves. He had some strong friends among the clergy, but nonetheless he was officially warned that he should neither teach nor advocate the Copernican system. It was a hard order to obey. As he bided his time, trying to appear resigned to the order from Rome, Galileo envisioned a major work, mustering all the evidence for and against the Copernican system. In 1630, after a decade of lobbying and cajoling, he obtained permission to publish his case, which appeared two years later as *Dialogue Concerning the Two Chief World Systems* (also known as *Dialogue on the Great World Systems*). It is perhaps his greatest work, and it was undeniably the most contentious.

Galileo's defense of Copernicus is cast in the form of a conversation among three men, one of whom defends the Aristotelian orthodoxy, one who argues for the new astronomy, and one who serves as an intermediary, considering the strong and weak points of the case. Though he had been admonished to give equal weight to both sides of the case, Galileo found it impossible to disguise his bias. The Aristotelian, for instance, is named Simplicio, and the current Pope Urban VIII, recognizing his own words in the mouth of an eponymous simpleton, couldn't help but take offense. A year after the publication of *Dialogue*, Galileo was summoned to Rome, accused of heresy, and, after a trial before the Inquisition, commanded to renounce any claims that the Earth moved or that the Sun was the center of the Universe. His book was banned, and he was placed under house arrest, where he remained for the remaining decade of his life.

In retrospect, of course, Galileo had it right, as Pope John Paul II recognized officially in 1992 when he expressed regret at the way the Church treated the whole affair. But despite the powerful evidence of the telescope, Galileo's enthusiasm for the new Sun-centered cosmology involved a leap of imagination that went far beyond the facts available at the time and far beyond what old-guard astronomers were prepared to accept.

Galileo was willing to take the leap in part because he suspected that there were new mathematical laws of physics waiting to be discovered. He spent much of the last years of his life writing about the laws of motion and the strength of materials, publishing his *Two New Sciences* in 1638. The work, far less argumentative than his book on cosmology, summarized a lifetime of studies in the behavior of matter. When he died in 1642, he had made major contributions to both the astronomy and physics that were to replace the celestial orbs and terrestrial elements of Aristotle.

Nearly another half-century, however, was to pass before the revolution in cosmology had made a full turn. By the late 1600s, the Sun-centered Universe was widely considered true, even though no one really knew how it worked. Not until Isaac Newton published his *Principia Mathematica*, or *Mathematical Principles of Natural Philosophy*, in 1687, could the matter be laid to rest. Newton showed how the behavior of everything in the Universe, from an apple to the moons of Jupiter, followed the same simple mathematical laws and how a single universal force of gravitation, acting between all bodies, governed the motions of the heavens. The telescope, to be sure, brought the heavens into view, but it took an equally powerful vision—the combined creative imaginations of Copernicus, Kepler, Galileo, and ultimately Newton—to bring it all into focus.

COSMOLOGY TODAY

Today we know that the Universe is a great web of dark matter that is marked by bright spots where there are galaxies. At places where two web filaments cross each other, there often is a cluster of hundreds or thousands of galaxies. A single galaxy in a cluster may contain several hundred billion stars. Many of these stars are likely to have planets, asteroids, and comets, just as the Sun does, and many of those planets must have moons and rings, like planets in our own solar system.

There are crucial uncertainties in our knowledge of cosmology—the structure and nature of the Universe—perhaps much more so than in any other branch of astronomy. To begin with, we don't know the nature of the dark matter, except that it consists of particles of a type or types that have yet to be identified and studied in the laboratory. It's the overwhelming majority of all matter in the Universe, it is dark and cannot be seen through our telescopes, and for all practical purposes, it is unknown to science. Dark matter interacts with ordinary matter in only one way that we have detected: through gravitational attraction.

There's an even greater unknown than dark matter. We observe that the Universe is expanding, as specified in the Big Bang theory. According to the Big Bang theory, everything began in a gigantic explosion of space itself, in which matter was created, at a point in time that we now date to 13.73 billion years ago. In fact, we can think of that as, for all practical purposes, the beginning of time. But the expansion of the Universe is not slowing down as had been expected. Such a slowdown, or deceleration, was the expected consequence of the mutual gravitational attraction of all matter in the Universe. Instead, we see that the expansion is accelerating, so that the rate at which the Universe is getting bigger is itself getting bigger. This acceleration that is driving the expansion of the Universe is attributed to so-called "dark energy." Matter and energy are equivalent according to Einstein's famous equation, $E = mc^2$, which relates the energy equivalent, or "rest energy" (E), of a particle of matter of mass (m) to the mass times the square of the velocity of light (c). Keeping this equivalence in mind and adding up all the mass and energy in the Universe, we find that the vast majority of all the mass energy in the Universe consists of dark energy. Dark energy is so powerful that it overrides the collective gravitation of the Universe. And we don't know what it is. We have come a long way since Galileo, but in the area of cosmology, we are still quite a bit in the dark. (No pun intended.)

In the Milky Way, there are over 300 known planetary systems beyond our own solar system and likely billions more waiting to be discovered. Everything that Galileo studied was in our solar system or in the Milky Way, which is just one of several hundred billion galaxies in the observable Universe. We are not in the center of the Milky Way, or even close, and we are certainly not in the center of the Universe, because it has no center. It goes on and on for billions of light-years in all directions and is about the same in all directions and at all distances, as far as we know. The parts of the Universe that we see in our deepest telescopic observations do look markedly different from the parts of the Universe close around us, but that's because when we look very far out in space, we can't see what the objects there look like now. We can only see what they looked like when the light focused by our telescope left those galaxies. And if a galaxy is billions of light-years away (and a light-year is the distance light travels through empty space over the course of a year), then we see the galaxy as it looked billions of years ago. The Universe is similar from place to place at the same time, but it has changed dramatically over time, as revealed by our most far-reaching telescopic observations.

The solar system, centered on the Sun, extends out about 700 billion miles to the outer limits of the Oort Cloud, the swarm of comets that represent the most distant objects that are gravitationally bound by the Sun, meaning that they are held in their orbits by the Sun's gravity.

As we discussed in Chapter 9, beyond the solar system, the Milky Way is a relatively large galaxy, with a flat, round galactic disk, about 100,000 light-years across, consisting of stars, star clusters, and interstellar gas and dust. At the galactic center, there is a huge black hole, with almost 4 million times the mass of the Sun, and surrounding the center there is a spherical mass of stars, the galactic bulge, and a much sparser spherical region, the galactic halo, 300,000 light-years in diameter. The halo contains globular star clusters as well as star streams, collections of stars with common origins and chemical composition that stretch over long distances. The star streams consist of the remains of small galaxies—dwarf galaxies—that were pulled apart by tidal force exerted by the Milky Way and are being absorbed into it.The Milky Way has about a dozen dwarf galaxies currently orbiting it as satellites, and close by there are one or two dwarf galaxies that may be recently arriving in our part of the Universe. These galaxies may just pass on by, or they may be subject to disruptive effects of the Milky Way.

Including the satellite galaxies of the Milky Way, there are a few dozen nearby galaxies that form a gravitationally bound system that is roughly 4 million light-years wide. This system, the Local Group of Galaxies, includes the Andromeda galaxy, which is about 2.9 million light-years away from Earth. The Andromeda galaxy is about the same size and shape as the Milky Way and has a flock of satellite dwarf galaxies of its own. It also has at least one star stream that shows that it has probably absorbed at least one such galaxy. Andromeda also has a giant black hole at the center.

There are many isolated galaxies in the Universe, as well as many groups of galaxies. However, there are much larger systems as well, the clusters of galaxies. A typical galaxy cluster may be 10 to 30 million light-years across and contain hundreds and in some cases thousands of galaxies. The nearest large cluster of galaxies is the Virgo cluster, about 54 million light-years away from the Earth and the Milky Way. It contains about 1,500 galaxies within a diameter of 8 million light-years, including some galaxies that are much larger, brighter, and more massive than the Milky Way.

The galaxies and clusters of galaxies are not spread evenly through the Universe. They are arranged in space along a so-called cosmic web of long filaments that cross each other in places called knots. And they are also arranged on sheets, or "walls," in space. These sheets surround immense regions where there are many fewer galaxies than on the sheets or in the filaments. So the arrangement of galaxies in space is reminiscent of the internal structure of a sponge. The relatively empty places between the walls and the filaments are called cosmic voids. The largest known cosmic void as of late 2007, the Eridanus void (named for the constellation where it is located), is almost a billion light-years across and is roughly 8 billion light-years from Earth. It was found in a survey of galaxies with the Very Large Array radio telescope in New Mexico, but as astronomers survey ever-larger regions of space at even greater distances, they may find bigger voids.

Clusters of galaxies are typically at the knots in the arrangement of filaments in the cosmic web, and in places, there are two or more distinct clusters near each other, making up so-called superclusters. However, though these superclusters are relatively close to each other in space, they are not gravitationally bound.

Pervading the vast reaches of space between clusters of galaxies and isolated groups of galaxies such as the Local Group there is a very thin gas called the warm-hot intergalactic medium, or, as it is whimsically named, the WHIM. Although the atoms in the WHIM are very far apart, it fills such a large volume of space that it contains the great majority of all normal matter in the Universe. ("Normal matter" is matter consisting of the atomic particles that are known to science, which, among other things, make up everything on Earth and all the objects that we can see in space.) The WHIM has more mass than all the hundreds of billions of galaxies put together.

So the great majority of mass energy in the Universe is present in the form of the unknown dark energy, the great majority of matter in the Universe is present in the form of the unknown dark matter, and the great majority of known forms of matter is present in the little-known and barely explored, virtually invisible WHIM. The actual objects that are present in space—the planets, stars, nebulae, galaxies, and so on—to which astronomers have devoted the great majority of their research over hundreds of years, are insignificant compared to the rest of the Universe and its matter, but they are all that we can see. This could be enough to discourage anyone from studying cosmology, or perhaps it's sufficient to make the curious, scientifically aware person read further.

Also pervading the entire Universe, reaching through the Milky Way and solar system and even down to the surface of the Earth, is a weak sea of radio waves, most prominent at microwave frequencies, known as the cosmic microwave background radiation. It was discovered in 1963 at Crawford Hill, New Jersey, by radio astronomers with the Bell Telephone Laboratories, who found an unexpected signal streaming into their antenna in what amounts to the farthest long-distance call ever received. It is a glow that was emitted by the hot matter that filled the Universe at an ancient time, about 370,000 years after the Big Bang, before the gas began to cool and slowly condense into stars and galaxies.

When the cosmic microwave background radiation is examined with high precision, as it was by the Cosmic Background Explorer (COBE) and later with even greater accuracy by the Wilkinson Microwave Anisotropy Probe (WMAP), two NASA satellites, it is found to exhibit very weak structure. That is, it differs from a perfectly even glow by the presence of areas on the sky that are only a few hundred thousandths brighter or a few hundred thousandths fainter than the average brightness. The giant Eridanus void, mentioned previously, is located at the same position on the sky as one of the relatively fainter regions in the background radiation recorded by the WMAP.

The slight differences, or "anisotropies," in the background radiation are crucial evidence that tells about the large-scale structure of the Universe, notably the cosmic web. Other evidence comes from the variety of galaxies and galaxy phenomena that we observe in the nearby Universe and other objects that we see in the distant Universe at much earlier times in the history of the Universe. But the earliest structure we detect in the history of the Universe and the farthest back that we see consists of the anisotropies in the background. They correspond to slight differences in the density of the hot primordial gas from the Big Bang, which are reflected in today's Universe by the large-scale structure of the cosmic web.

In the Milky Way and many other galaxies in the nearby Universe, we see spectacular explosions of massive stars as well as explosions in which a small star in a binary star system explodes with such great force that the entire star is shattered. These two kinds of supernova explosions produce high-speed expanding nebulae (supernova remnants) that drive outward into the interstellar gas of the galaxies in which they occur.

At the centers of the Milky Way, Andromeda, and most other nearby galaxies, and more distant galaxies to the extent our telescopic powers allow, we find evidence for the presence of supermassive black holes with masses that are millions and even several billions times the mass of the Sun. For example, when we measure the velocities of stars rotating very close to the center of a galaxy and discover that those speeds drop off for stars orbiting at slightly larger distances, we can infer the presence of an unseen gravitating body in which a very great mass is concentrated in a small radius—in other words, a supermassive black hole. Typically there is one supermassive black hole at the center of each galaxy, but in rare cases there are two, suggesting that the galaxy may have swallowed another galaxy, which was once centered on one of the two black holes.

In the nearby Universe, we see much evidence of galaxies merging, colliding, or pulling one another apart. And in the very distant Universe, at the limits of observation of the Hubble Space Telescope, we see many galaxies in early stages of assembly, still much smaller in that early epoch than the mature galaxies of the Universe today.

A large galaxy can grow by mergers. The Milky Way and Andromeda galaxies increase in this way by swallowing some of their satellites. Galaxies of similar size that pass each other in space can interact as well when their mutual gravitational attraction pulls long streams of interstellar gas from each other. New stars form in the streams, producing striking "tidal tails" in such interacting galaxy pairs as the Antennae galaxies NGC 4038 and 4039, located about 63 million light-years from Earth in the constellation Corvus. (NGC stands for "New General Catalogue," which is actually an old catalogue of galaxies, star clusters, and nebulae, published in 1888, when the distinction between nebulae as gas clouds and galaxies as much larger objects consisting of stars and gas was unknown.) Images made with the Hubble Space Telescope show that the collision of the Antennae galaxies triggered the formation of more than 1,000 star clusters.

Galaxies sometimes suffer head-on collisions, with one passing right through the other. Stars in a given galaxy are so far apart that few actually collide with stars from the other galaxy, but interstellar gas clouds are strongly affected. When one galaxy passes through the approximate center of another, the more massive one may produce a so-called ring galaxy. At its center is the dense, bright stellar core of the target galaxy, and at a large radius from the center, there is a great round hoop of gas that was stripped from the smaller galaxy and in which many new stars form. The NGC 4650A galaxy, about 140 million light-years from Earth in the constellation Centaurus, is a good example. Systems with tidal tails and ring galaxies are observed in the local Universe; they are phenomena that are occurring today and may have also occurred long ago.

The Antennae galaxy and the ring galaxy NGC 4650A are considered part of the local Universe, so that they illustrate the kinds of objects and phenomena that characterize the Universe at the present time, along with relatively normal disk and elliptical galaxies, dwarf galaxies, and others. However, in the very distant Universe, seen in the deepest available telescopic images made with the Hubble Space Telescope, the Hubble Deep Field, the Hubble Ultra Deep Field, and some others, we find a different situation. Back then, while some galaxies are not clearly very different from galaxies today (as far as we can tell), a great many galaxies look irregular or unusual, and a larger fraction of galaxies seem to be colliding (although these collisions occurred long ago). Many of these distant young galaxies have a chunky appearance. Small galaxies or building blocks of galaxies are coming together to form larger systems.

Large galaxies found today are usually one of two types. They are disk galaxies, like Andromeda and the Milky Way, which have much interstellar gas and many young stars and often show spiral patterns as well as a central bulge of stars. Or they are elliptical galaxies, which may be great spherical clouds of stars like M87, the giant galaxy in the Virgo cluster, or they may be shaped like a football. Elliptical galaxies don't have disks. They are all bulge, and they don't have cool interstellar gas clouds where stars can form. However, in some cases they contain large clouds of very thin and hot gas, which are confined by gravity much stronger than expected from the visible appearance of the galaxy. That is, the combined mass of all the stars in the elliptical galaxy is much less than would be required to keep the hot clouds, which are seen in X-ray images made by satellites such as the Chandra X-ray Observatory, confined to their present volume. About 20 times as much mass as is present in stars is present in the elliptical galaxies. It is dark matter.

Whether a galaxy is a disk system or an elliptical, it has a super-massive black hole at the center, and the mass of the black hole is correlated with the mass of the galaxy bulge in which it is located. The bigger the bulge, the bigger the black hole.The most massive black holes are at the centers of the most massive galaxies, which are giant ellipticals like M87.

Many galaxies contain "active galactic nuclei," meaning that they each have a very bright object at their galactic center, or nucleus. When it is especially bright—sometimes as bright or brighter than the entire remainder of its home, or "host," galaxy—the bright central object is called a quasar. Active galactic nuclei often send bright jets of matter streaming into space in opposite directions. M87 is famous for a bright jet that was discovered years before astronomers knew of active galactic nuclei or found the first quasar. Until then, the jet was just an unexplained curiosity.

Thanks to the Hubble Space Telescope, we know that there is supermassive black hole at the center of M87 which sucks in gas and perhaps even stars from the galaxy around it. The matter falling toward the black hole assumes a flattened distribution, the accretion disk, in which it spirals inward. Eventually, much of it falls into the black hole, which grows with time as long as more matter is falling in. However, some of the matter from the accretion disk is thrust outward perpendicularly to the disk, producing the powerful jets seen in M87 and other active galaxies. In M87, one jet is aimed roughly in our direction and the other in the opposite way. The jet of matter coming toward us moves at a significant fraction of the velocity of light and always looks much brighter than the other jet from our perspective. In some active galaxies, the jets feed great "radio lobes," blobs of radio-emitting gas detected by radio telescopes but not seen in photographs made in visible light. The blobs are usually far beyond the visible extent of the active galaxy.

Another type of unusual galaxy is observed throughout the Universe, although the extreme cases are the farthest away. These objects, called starburst galaxies, are caught in an episode of rapid star formation in which many stars are born at once at a much higher rate in new stars per year or total mass of new stars per year than in a normal galaxy, or in the same galaxy (as we infer) at earlier and later times. In the most extreme starburst galaxies, huge masses of interstellar dust glow with infrared light from the heat of the hot young stars within the dust clouds. These galaxies exhibit strong winds, which blow interstellar matter from within the galaxy into intergalactic space. The matter expelled in this way is not primordial gas from the Big Bang; it is gas that has been enriched by nuclear processes that occurred within stars and was then expelled into the interstellar medium through winds from red giant stars and through supernova explosions. From the interstellar medium, such matter eventually makes its way into intergalactic space, and together with the gas expelled from active galaxies by jets from the vicinity of their central black holes, it contributes to the WHIM. More importantly, it deprives the source galaxies of some of their supply of material in which new stars can form.

A galaxy spends only a small part of its life as an active galaxy, in which matter falls into the central black hole and produces jets, X-rays, radio emission, and sometimes gamma rays. (Some jets may also produce high-energy cosmic rays, which are protons and other subatomic particles that travel at a high fraction of the speed of light.) And it spends only a small fraction of its life as a starburst galaxy. In some cases, it is both at once. There were more quasars and especially intense starburst galaxies in the distant past than in the present Universe.

The modern view of a galaxy is dependent on dark matter. There is abundant evidence that the great majority of matter in almost every galaxy is the unidentified dark matter that also makes up the cosmic web. For example, if the only matter in a galaxy was visible matter, then stars that orbit at the outskirts of the galaxy would orbit the center at much slower speeds than stars inward of them, just as the Earth goes around the Sun at a slower speed than the inner planets Mercury and Venus, and the Earth in turn orbits the Sun at a faster speed than the outer planets Mars, Jupiter, Saturn, and so on. But numerous observations, first performed by the astronomer Vera Rubin, reveal that at a certain large distance from the center of a galactic disk, the speeds level off rather than drop off for stars farther from the center. This is attributed to a vast halo of dark matter, much larger than the visible galactic disk. In this halo, the matter is much less concentrated toward the center than is true of the stars in the galaxy. As in elliptical galaxies, the dark matter in a disk galaxy is about 20 times as massive as the ordinary matter of stars.

Dark matter was originally detected in clusters of galaxies by the astronomer Fred Zwicky, although few if any other scientists believed him. He found that galaxies moving through their clusters are orbiting at speeds much too fast to be explained by the total amount of mass visible in the cluster, as estimated by adding up the masses of all the galaxies. The greater the total mass, the faster galaxies orbit, just as the speed at which the Earth orbits the Sun depends on the mass of the Sun. (If the mass of the Sun was less than it actually is, it would exert less force on the Earth, and our planet would orbit less rapidly than it does now, making the year longer than its present 365¼ days.) Unless invisible matter was present in the cluster in much greater amounts than the visible matter of the galaxies, the cluster would break up, since galaxies would not be bound to the system and would fly away into intergalactic space.

Dark matter present in galaxies and in clusters of galaxies is also detected, and its degree of concentration toward their centers is measured, through observations of gravitational lensing. This is a phenomenon in which the appearance of a distant galaxy as seen from Earth is distorted by the action of the gravity of a massive galaxy or a cluster located roughly along the line of sight to the distant object. In some cases, in which the object at intermediate distance (called the "gravitational lens") is almost perfectly lined up with the distant galaxy or quasar (the "lensed object"), the lensed object doesn't look like a galaxy at all. It appears to us a luminous circle, a so-called Einstein ring. Astronomers located elsewhere in space, with no gravitational lens between them and the same galaxy, would not see the ring; they would just see an ordinary galaxy. The extent of the gravitational lensing is much greater than can be explained by the mass of visible objects in the lensing galaxy or lensing cluster, and is accounted for by the presence of dark matter that is much more massive than the visible matter.

According to our present understanding, all the primordial matter of the Universe was created in an infinitesimal space in the Big Bang 13.73 billion years ago. It was not an explosion in space but the explosion *of* space that formed our Universe, which continues to expand. Infinitesimal fluctuations in the density of matter in the earliest instants of time, when for a fraction of a second there was an extremely rapid expansion, or "inflation," led to the condensation of huge clouds of dark matter, whose gravity brought along the ordinary matter that was and is present to a much smaller degree. The matter simply flowed toward the denser regions, whose gravitational attraction was greater than average. The flows of condensing gas generated sound waves in the gaseous medium of the Universe, so-called acoustic oscillations, which correspond to the fluctuations detected in the background radiation by the COBE and WMAP satellites. These fluctuations produced the pattern of the cosmic web, which controls the locations at which galaxies and clusters of galaxies formed. In fact, a pattern in the statistical distribution of galaxies across the sky that corresponds with the acoustic oscillations was detected by astronomers working with data from the Sloan Digital Sky Survey, which maps huge regions of space in three dimensions with a telescope in New Mexico. From initial infinitesimal differences in density in the Big Bang, this pattern has grown as the Universe expanded.

With time, galaxies exhaust their content of interstellar gas from which new stars can form. Often, their mergers with other galaxies lead to an end state in which there is a large elliptical galaxy with no disk of gas or other cool interstellar medium from which new stars can form.

Little is known about the dark energy that causes the expansion of the Universe to become ever faster with time. A crucial issue is whether the nature of the dark energy is such that the expansion will increase with essentially no limit, eventually causing even individual galaxies like our Milky Way and planetary systems like the solar system to be torn apart. This dire outcome is not certain, but it is a possibility. Astronomers have initiated a variety of projects with huge ground-based telescopes and proposed future space telescopes that should be capable of settling the issue.

EPILOGUE

Four centuries after Galileo, astronomers and their telescopes have explored the solar system, discovered many other planetary systems, mapped the Milky Way galaxy, and found once inconceivable cosmic objects such as giant black holes. They have learned what powers the stars and how they are born, evolve, and die. They are beginning to unravel the mysteries of how, why, and when galaxies formed and how they grow, and to a fair degree, they have determined the structure of the Universe and how it most likely took that form. Yet the mysteries that remain are as great as those that faced scientists when Galileo announced his first telescopic findings in the early seventeenth century.

We have some clues but almost no solid information about the nature of the vast majority of matter in the Universe, the dark matter, and even less data on the identity and properties of the phenomenon that constitutes almost 75 percent of the mass energy in the Universe, dark energy. Telescopes reveal the presence of dark matter throughout the Universe, but perhaps they may never tell us what dark matter is. The answer instead could come from any of several laboratory experiments now underway on Earth, such as the Cryogenic Dark Matter Search that physicists are conducting in an old iron mine a half-mile beneath the surface of the Earth in northern

Minnesota. More likely, it will come from future experiments yet to be initiated. Like dark matter, dark energy is as much a problem for physicists as astronomers, but crucial information on its nature will almost surely come from future telescopes on Earth or in space, or perhaps even from existing telescopes that are being instrumented and applied in new ways. It's not likely that dark energy will be detected and identified in a laboratory experiment. It seems that major advances in telescope technology may not be required for this research; it may simply require a great deal of sophisticated data gathering and data processing.

Many scientists believe that we will not have a proper understanding of how the Universe works until there is a well-founded theory that unifies the theory of gravity with the physical laws that govern matter at the smallest scales within the atom. Einstein himself failed in this attempt, and hundreds of excellent physicists who have followed him have had little more success. The answer may come someday from laboratory experiments that explore the force exerted by gravity over extremely small distances, or somehow from astronomical observations that reveal how objects move under the influence of gravity over very great distances in space. Some think the answer may lie in the presence of higher dimensions beyond the three spatial dimensions and the time dimension that we know of so far.

The detection of waves of gravitational radiation from objects in the cosmos is one of the prime goals of astronomers and physicists and could occur in the next decade or two. It's predicted by Einstein's general theory of relativity. The fabric of space is briefly deformed as a gravitational wave passes, just as a swimmer bobs up and down on the ocean as a water wave comes along. Gravitational wave observatories in the United States, Italy, and Germany are joining hands in the search, and there's an extremely sensitive facility under design that would be located in space through the joint efforts of NASA and the European Space Agency.

We strongly suspect, but have absolutely no evidence, that life is present elsewhere in the Universe, probably including intelligent life. We don't think that we are alone, but we have no communications with other civilizations among the stars. Proclaimed "evidence" of fossil microbial life on Mars, discovered in a single rock that was knocked off of the planet, was announced by NASA in 1996. However, that evidence turns out to be inconclusive at best.

Future planetary probes will land on Mars, conduct robotic experiments more sophisticated than those performed by earlier probes, and perhaps even return samples of rock and soil. If they find fossil or even presently existing life, the discoveries will have an enormous impact on the thinking of every scientifically inclined person. But if they find no evidence, it will mean little to the issue of whether there is or has been life on Mars. The absence of evidence is not evidence of absence. It's just a lack of information.

Life may exist in the underground oceans of one or more of the moons of Jupiter discovered by Galileo himself, Europa and Ganymede. It may exist on Saturn's moon Enceladus, below the surface of the south polar region where geysers spout.

Telescopes won't find life on Mars, Europa, Enceladus, or anywhere else in our solar system; nor will they rule it out. But instrumented space probes might find it someday, exploring locations revealed by telescopes on earlier spacecraft.

Radio telescopes now existing, like the 1,000-foot dish at Arecibo, Puerto Rico, might detect radio signals from extraterrestrial civilizations residing on planets orbiting stars in the Milky Way. More likely, if such signals are detected, they will be captured by future huge radio telescopes, like the Square Kilometre Array, whose design is under study at the National Radio Astronomy Observatory in the United States and at astronomical institutes abroad.

Advances in the Internet and like methods of global communications allow telescopes at separate locations to be tied together in networks that will someday facilitate astronomical observations. These combined observations may be more powerful than the data from the largest telescopes now on the drawing boards, like the European Extremely Large Telescope, which might be 140 feet or more in diameter. Indeed the networks could incorporate these largest telescopes.

Increasingly, astronomers use modern telescopes on the ground much like telescopes in space. They not only don't look through the telescopes, they often don't operate them and may not even be present when the telescopes make the desired observations. The astronomers fill out scripts that specify the details of the observations that they need, which are executed by computers under the control of observatory specialists. The end products are digital data transmitted to the researchers.

Telescopes in space will move outward beyond low Earth orbit where the Hubble Space Telescope circles every 90 minutes or so. That's already the plan for the James Webb Space Telescope, the much larger successor to the Hubble. The JWST will be put in orbit around an empty point in space, called L2 (second Lagrangian point), almost 1 million miles from Earth in the direction opposite the Sun. At that location, the gravitational attractions exerted on the spacecraft by the Sun and the Earth are equal. The telescope will search for the first galaxies formed in the Universe and help to answer the question of how they formed and what they were like.

Already now, and to a greater extent in the future, astronomers are data miners who strive to solve celestial mysteries by sifting through large archives of digital data from telescopes. In these cases, the telescope is not directed toward a specific star or galaxy but surveys millions of stars and galaxies, producing databases that can be mined by many different investigators who seek to answer a variety of questions.

Telescopic observations were much simpler for Galileo, but their consequences were at least as profound as those of nearly all observations that astronomers are making today.

BIBLIOGRAPHY

SELECTED READINGS ON GALILEO AND TELESCOPES

Andersen, Geoff. *The Telescope: Its History, Technology, and Future.* Princeton, NJ: Princeton University Press, 2006.

Bell, Louis. *The Telescope.* Mineola, NY: Dover Publications, 1981.

Drake, Stillman. *Galileo at Work: His Scientific Biography.* New York: Dover Publications, 1978.

Frova, Andrea, and Mariapiera Marenzana. *Thus Spoke Galileo: The Great Scientist's Ideas and their Relevance to the Present Day.* New York: Oxford University Press, 2006.

Galilei, Galileo. *Dialogue Concerning the Two Chief World Systems.* Translated by Stillman Drake. Berkeley, CA: University of California Press, 1953.

Galilei, Galileo. *Discoveries and Opinions of Galileo.* Translated by Stillman Drake. New York: Anchor, 1957.

Galilei, Galileo. *Sidereus Nuncius, or The Sidereal Messenger.* Translated by Albert Van Helden. Chicago: University of Chicago Press, 1989.

Grant, Edward. *Planets, Stars, & Orbs: The Medieval Cosmos, 1200–1687.* Cambridge: Cambridge University Press, 1994.

King, Henry C. *The History of the Telescope*. New York: Dover Publications, 1955.

Koestler, Arthur. *The Sleepwalkers: A History of Man's Changing Vision of the Universe*. New York: Macmillan, 1959.

Lowe, Jonathan. "Next Light: Tomorrow's Monster Telescopes." *Sky & Telescope* 115, no. 4 (2008): 20–25.

Sharratt, Michael. *Galileo: Decisive Innovator*. Cambridge: Cambridge University Press, 1994.

Van Helden, Albert. *Measuring the Universe*. Chicago: University of Chicago Press, 1985.

Van Helden, Albert, and Elizabeth Burr. *The Galileo Project*. http://galileo.rice.edu/.

SELECTED READINGS ON ASTRONOMY TODAY

THE MOON

Hartmann, William K. *Moons & Planets*. 5th ed. Belmont, CA: Brooks/Cole, 2005.

Heiken, Grant H., David T. Vaniman, and Bevan M. French, eds. *Lunar Sourcebook*. Cambridge: Cambridge University Press, 1991.

Jolliff, Bradley L., Mark A. Wieczorek, Charles K. Shearer, and Clive R. Neal, eds. *New Views of the Moon*. *Reviews in Mineralogy & Geochemistry* 60. Chantilly, VA: Mineralogical Society of America, 2006.

Spudis, Paul D. "The Moon." In *The New Solar System*, 4th ed., edited by J. Kelly Beatty, Carolyn Collins Petersen, and Andrew Chaikin, 125–140. Cambridge, MA: Sky Publishing Corporation, 1999.

Spudis, Paul D. *The Once and Future Moon*. Washington, DC: Smithsonian Institution Press, 1998.

THE SUN

Carlowicz, Michael J., and Ramon E. Lopez. *Storms from the Sun: The Emerging Science of Space Weather*. Washington, DC: Joseph Henry Press, 2002.

Lang, Kenneth R. *The Cambridge Encyclopedia of the Sun*. Cambridge: Cambridge University Press, 2001.

Odenwald, Sten. *The 23rd Cycle: Learning to Live with a Stormy Star.* New York: Columbia University Press, 2001.

Science. Special issue on Ulysses. Vol. 268, no. 5213 (1995).

JUPITER

Beebe, Reta. *Jupiter: The Giant Planet.* 2nd ed. Washington, DC: Smithsonian Institution Press, 1997.

Davies, Ashley G. "Volcanism on Io: The View from Galileo." *Astronomy & Geophysics* 42, no. 2 (2001): 10–15.

de Pater, Imke, and Jack J. Lissauer. *Planetary Sciences.* Cambridge: Cambridge University Press, 2001.

Greenberg, Richard. *Europa: The Ocean Moon.* Berlin: Springer-Verlag, 2005.

Johnson, Torrence V. "A Look at the Galilean Satellites After the Galileo Mission." *Physics Today* 57, no. 4 (2004): 77–83.

Kerr, Richard A. "Jupiter's Two-Faced Moon, Ganymede, Falling Into Line." *Science* 291, no. 5501 (2001): 22–23.

West, Robert A. "Atmospheres of the Giant Planets." In *Encyclopedia of the Solar System*, 2nd ed., edited by Lucy-Ann McFadden, Paul R. Weissman, and Torrence V. Johnson, 383–402. San Diego: Academic Press, 2007.

SATURN AND BEYOND

Coustenis, Athena. "What Cassini-Huygens Has Revealed About Titan." *Astronomy & Geophysics* 48, no. 2 (2007): 14–20.

Durisen, Richard H. "Planetary Rings: Moonlets in a Cosmic Sandblaster." *Mercury* 28, no. 5 (1999): 10–23.

Hubbard, William B. "Interiors of the Giant Planets." In *The New Solar System*, 4th ed., edited by J. Kelly Beatty, Carolyn Collins Petersen, and Andrew Chaikin, 193–200. Cambridge, MA: Sky Publishing Corporation, 1999.

Ingersoll, Andrew P. "Atmospheres of the Giant Planets." In *The New Solar System*, 4th ed., edited by J. Kelly Beatty, Carolyn Collins Petersen, and Andrew Chaikin, 201–20. Cambridge, MA: Sky Publishing Corporation, 1999.

Kerr, Richard A. "How Saturn's Icy Moons Get a (Geologic) Life." *Science* 311, no. 5757 (2006): 29.

Lancaster, Nicholas. "Linear Dunes on Titan." *Science* 312, no. 5774 (2006): 702–703.

Lopes, Rosaly M. C., Karl L. Mitchell, Stephen D. Wall, Giuseppe Mitri, Michael Janssen, Steven Ostro, Randolph L. Kirk, et al. "The Lakes and Seas of Titan." *EOS* 88, no. 51 (2007): 569–570.

Porco, Carolyn C., P. C. Thomas, J. W. Weiss, and D. C. Richardson. "Saturn's Small Inner Satellites: Clues to Their Origins." *Science* 318, no. 5856 (2007): 1602–1607.

Porco, Carolyn. *CICLOPS: Cassini Imaging Central Laboratory for Operations.* http://ciclops.org/index.php.

Science. Special issue on Cassini at Saturn. Vol. 307, no. 5713 (2005).

Venus

Cattermole, Peter. *Venus: The Geological Story.* Baltimore: Johns Hopkins University Press, 1996.

Grinspoon, David H. *Venus Revealed: A New Look Below the Clouds of Our Mysterious Twin Planet.* Reading, MA: Helix Books, 1997.

Robertson, Donald F. "Parched Planet." *Sky & Telescope* 115, no. 4 (2008): 26–31.

Saunders, R. Stephen. "Venus." In *The New Solar System*, 4th ed., edited by J. Kelly Beatty, Carolyn Collins Petersen, and Andrew Chaikin, 97–110. Cambridge, MA: Sky Publishing Corporation, 1999.

Special issue on Venus Express, *Nature* 450, no. 7170 (2007).

Comets

Brandt, John C., and Robert D. Chapman. *Introduction to Comets.* 2nd ed. Cambridge: Cambridge University Press, 2004.

de Pater, Imke, and Jack J. Lissauer. *Planetary Sciences.* Cambridge: Cambridge University Press, 2001.

Festou, M. C., H. U. Keller, and H. A. Weaver, eds. *Comets II.* Tucson, AZ: University of Arizona Press, 2004.

Lazzaro, Daniela, Sylvia Ferraz-Mello, and Julio Angel Fernández, eds. *Asteroids, Comets, and Meteors.* Cambridge: Cambridge University Press, 2006.

Science. Special issue on Stardust. Vol. 314, no. 5806 (2006).

Yeomans, Donald K. *Comets: A Chronological History of Observation, Science, Myth, and Folklore.* New York: John Wiley & Sons, 1991.

The Stars and the Milky Way

Bennett, Jeffrey O., Donahue, Megan, Schneider, Nicholas, and Voit, Mark. *The Cosmic Perspective*. 5th ed. Benjamin Cummings, 2007.

Cowen, Ron. "The Milky Way's Middle: Getting a Clear View." *Science News* 161, no. 8 (2002): 122.

Harwit, Martin. *Cosmic Discovery: The Search, Scope, and Heritage of Astronomy*. Cambridge, MA: MIT Press, 1984.

Hester, Jeff, David Burstein, George Blumenthal, Ronald Greeley, Bradford Smith, Howard Voss, and Gary Wegner. *21st Century Astronomy*. 2nd ed. New York: W. W. Norton & Company, 2007.

Kormendy, John. "Galactic Rotation in Real Time." *Nature* 407, no. 6802 (2000): 307–309.

Petersen, Carolyn Collins, and John C. Brandt. *Hubble Vision: Astronomy with the Hubble Space Telescope*. Cambridge: Cambridge University Press, 1995.

Thorne, Kip S. *Black Holes and Time Warps: Einstein's Outrageous Legacy*. New York: W. W. Norton & Company, 1995.

van den Heuvel, Edward P. J. "Astrophysics: A Story of Singular Degeneracy." *Nature* 451, no. 7180 (2008): 775–77.

Cosmology

Freedman, Wendy L., ed. *Measuring and Modeling the Universe*. Cambridge, MA: Cambridge University Press, 2004.

Goldsmith, Donald. *The Runaway Universe: The Race to Find the Future of the Cosmos*. New York: Basic Books, 2000.

Ho, Luis C., ed. *Coevolution of Black Holes and Galaxies*. Cambridge: Cambridge University Press, 2004.

Kirshner, Robert P. *The Extravagant Universe: Exploding Stars, Dark Energy, and the Accelerating Cosmos*. Princeton, NJ: Princeton University Press, 2005.

Mather, John C., and John Boslough. *The Very First Light: The True Inside Story of the Scientific Journey Back to the Dawn of the Universe*. New York: Basic Books, 2008.

Silk, Joseph. "The Dark Side of the Universe." *Astronomy & Geophysics* 48, no. 2 (2007): 30–38.

Science. Special issue on the Cosmic Web. Vol. 319, no. 5859 (2008).

ABOUT THE AUTHORS

Dr. Stephen P. Maran spent more than 35 years in NASA, working on the Hubble Space Telescope and other scientific projects and is the press officer for the American Astronomical Society. His 10 previous books include *Astronomy for Dummies*® and *The Astronomy and Astrophysics Encyclopedia*. His awards and honors include the naming of an asteroid for him by the International Astronomical Union, the NASA Medal for Exceptional Achievement, the George Van Biesbroeck Prize of the American Astronomical Society and the Astronomical Society of the Pacific's Klumpke-Roberts Award for outstanding contributions to the public understanding and appreciation of astronomy.

Laurence Marschall, PhD, is the W.K.T. Sahm Professor of Physics at Gettysburg College where he teaches courses in astronomy, physics and science writing. He writes a regular column on science books of note for *Natural History* magazine and serves as deputy press officer of the American Astronomical Society. In addition to more than 40 articles in professional journals, Marschall has written for publications such as *Sky and Telescope, Astronomy, Natural History, Discover, Harper's, Newsday* and *The New York Times Book Review*. His book *The Supernova Story* (Princeton Science Library, 1994) has been widely praised for its readability. He was awarded the 2005 Education Prize of the American Astronomical Society for his work in furthering undergraduate instruction in astronomy.